The Environment,
Our Natural Resources,
and Modern Technology

The Environment,
Our Natural Resources,
and Modern Technology

THOMAS R.
DEGREGORI

Iowa State Press
A Blackwell Publishing Company

THOMAS R. DEGREGORI is a professor of economics, University of Houston, Texas. DeGregori received a BA in Philosophy and Government (1959) and an MA in Economics from the University of New Mexico, and a Ph.D. in Economics from the University of Texas. He has been involved in development work in the field and as a lecturer, teacher, and policy advisor in about 50 countries in Africa, Asia, and the Caribbean, and has served as a policy advisor to regional, national, and international donor organizations.

© 2002 Iowa State Press
A Blackwell Publishing Company
All rights reserved

Iowa State Press
2121 State Avenue, Ames, Iowa 50014

Orders: 1-800-862-6657
Office: 1-515-292-0140
Fax: 1-515-292-3348
Web site: www.iowastatepress.com

Authorization to photocopy items for internal or personal use, or the internal or personal use of specific clients, is granted by Iowa State Press, provided that the base fee of $.10 per copy is paid directly to the Copyright Clearance Center, 222 Rosewood Drive, Danvers, MA 01923. For those organizations that have been granted a photocopy license by CCC, a separate system of payments has been arranged. The fee code for users of the Transactional Reporting Service is 0-8138-0869-3 (hardcover); 0-8138-0923-1 (paperback)/2002 $.10.

First edition, 2002

Library of Congress Cataloging-in-Publication Data

DeGregori, Thomas R.
  The environment, our natural resources, and modern technology / by
Thomas R. DeGregori.—1st ed.
     p. cm.
Includes bibliographical references and index.
  ISBN 0-8138-0869-3 (hardcover); 0-8138-0923-1 (paperback)
  1. Environmentalism 2. Human ecology. I. Title.
  GE195 .D446 2002
  304.2—dc21
                        2002000057

The last digit is the print number: 9 8 7 6 5 4 3 2 1

Contents

Preface

A brief note on sources is in order. This book could not have been written without the Internet and links from posted news articles to original sources. Being able to go from a news article to the peer-reviewed scientific study offered an incredible array of diverse sources, including access to peer-reviewed specialist journals. In preparing this manuscript, I felt it necessary to remove numerous secondary sources to keep the reader from being overwhelmed by citations. My editor wisely culled further. In two instances, I had the secondary source (or sources) restored where I had been unable to locate the quote or other reference in the original. The reader can be assured then that I have read the entire source cited even though I may have encountered it through a more popular article.

The reader should find more than adequate documentation for all issues discussed. On my home page (http://www.uh.edu/~trdegreg), I have posted a supplementary bibliography, which is combined with the one that I posted for my recently published book, *Agriculture and Modern Technology: A Defense*. Other recent articles of mine are also posted there.

Using the Internet for research also adds some complexities. The traditional methods of referencing do not always mesh with electronic technology. When one quotes from an online posting of a printed article, it is generally not possible to give the page number(s) unless it is in PDF format. For those who go online to check my sources, this is no problem; simply search the article using a couple of words from the quote. Those who go for the hard copy may have some difficulty finding the quote and its context. Unless the cited article is available only online, the bibliography follows tradition, refers to the hard copy, and does not provide the URL (which for some of my regular sources changed over the course of my research), though I may have read it online. Some sources changed names—Nando Net, Nando Media, and Nando Times, for example. I followed the name used at the time of the posting. In most cases, a search engine will still find the journal, the online web page, or other source.

In addition to the Internet, satellites have allowed the same newspaper to be printed in numerous locations around the world and around the United States in different regionalized editions, making some of the finest news sources truly global publications. The *International Herald Tribune* is jointly owned by the *New York Times* and the *Washington Post,* and some of their articles appear in editions around the world many days or even a week or more later than in the home edition. In Houston, I receive the *New York Times* and the *Financial Times* of London every day on the day of publication. When I went online for the *Financial Times* series on NGOs, I noticed that the exact same articles had not only different dates but also different titles. In citing them, I use the date and title for the edition that I was reading, which means that the article may well have appeared on a different date with a different title in another edition. This is normally only one day off, but I have noted that for some publications there may be a difference of several days in the date of stories between a European and American edition; occasionally it will appear in only one. Again, this is not a problem if one goes online, but it might be if one is searching an archival newspaper or microfilm.

I have been fortunate to be able to lead an extraordinarily fascinating and extremely happy life. There are many who have positively touched my life and from whom I have learned—my parents, my sister, extended kin, childhood friends, and teachers on through to my adult years with editors, authors of works that I have read, family, colleagues, students, and friends in Houston and around the world. There are far too many categories of people to begin to mention, let alone express my profound indebtedness to individuals or groups.

I dedicate this book to all who have touched my life, whenever and however they may have done so; they have made my life and work what it is and will, I trust, continue to be. As is customary and appropriate at this point, I acknowledge that the work is of my hand and the errors are mine alone.

Thomas R. DeGregori, Ph.D.
Professor of Economics
University of Houston
Department of Economics
204 McElhinney Hall
Houston, Texas 77204-5019
Ph. 001-1-713 743-3838
Fax 001-1-713 743-3798
Email trdegreg@uh.edu
Web homepage http://www.uh.edu/~trdegreg

Introduction

In a recently published book, I begin with the following lines, variations of which I also use in lectures and other presentations. "This book deals with one of the great paradoxes of the 20th century. It was a century characterized by economic and technological change of unprecedented rapidity as shown by all of our economic indicators; an increasing number of economists now argue that they substantially understate the magnitude of these gains (Nordhaus 1997; DeLong 1991–2000). The noneconomic indicators are just as spectacular if one looks at life expectancy, health and the increase in per capita food supply that more than accommodates an increase in population that virtually all 'experts' believed could not be fed" (DeGregori 2001).

To repeat once again, the century that closed the second millennium was one in which there was economic and technological change of unprecedented rapidity. By every measure of well-being—economic, health, life expectancy, infant mortality, per capita food supply, etc., the human condition has never progressed as fast and has never been better. "Medical advances have not only lengthened life expectancy, but have also reduced its variance in the developed world," which means the poor in developed countries have improved both their absolute and relative well-being (May 2001, 891). Fogel and others have shown that there is a direct correlation between height and longevity. In the early 19th century, "a typical British male worker at maturity was about five inches shorter than a mature male of upper-class birth," a gap that has been reduced to about an inch today, which reflects the narrowing of the life expectancy gap also (Fogel 2000, 143–144). And one respected scientist has challenged the notion that the global gap between the rich and the poor has been widening

(Castles 2000). Life expectancies have definitely been increasing faster in developing countries than in the developed, though the AIDS crisis in Africa is reversing this trend for far too many people. As I have often stated elsewhere, the paradox is that these gains are largely denied, if only by implication, and the science and technology that allowed them to happen have been under attack for almost the entire century. The alleged dangers of modern life are accepted as unchallenged and unchallengeable truths by large segments of the population who are also largely the beneficiaries of this century of extraordinary change. It is our contention that this belief system is a manifestation of an antitechnology elitism. Further, the commonalities of the various myths about the virtues of the "primitive" or of the "natural," which we explore in this book, are not accidental but flow from an underlying antitechnology, antiscience mind-set.

In the book from which the opening quote is taken, I deal primarily with issues of agriculture and food supply, as well as health, medical practice, and life expectancy. Here the focus is on the consumption practices that reflect the phobias and beliefs that deny and/or reject the technological and scientific transformations that have given us longer, healthier lives. Denial, in the form of assertions that, for example, "chemicals" are killing us, gives rise to beliefs that the demon technology can be exorcised by more "natural" lifestyles that include more "all natural" organic foods and herbs, "alternative" medicine and "holistic" healing, and lifestyles that bring one ever closer to nature. As we argue in Chapter 1, only elitist affluent societies can give rise to "green consumerism" and related consumption and leisure activities such as ecotourism.

Being affluent and being "ecologically correct" would appear to be a contradiction in terms to many. Whether it is from a sense of guilt for being affluent, or from some other affectation, some have turned to socially responsible consumption, or "green consumerism." However good it may make the participants feel, even the most cursory examination of the prices paid for these items of "green consumption" shows them to be beyond the means of all but a privileged few. And to the extent economists are correct that prices measure scarcity, then any attempt to expand the consumption of these "green" items would likely increase their price even further. "Saving the planet" seems to be an endeavor that only the rich can truly afford. To the nonparticipants in "green consumerism," some of the lengths to which it is carried seem more superstitious than substantial and even to some extent ridiculous.

A sampling of such is not meant to offend the believers, but to reflect a rather significant difference in perception between a subgroup and the larger community.

To the extent that "green consumerism" is simply a fetish of the affluent, the rest of us can view it with a detached amusement; otherwise this form of consumption would be none of our business. But as we repeat over and over again in this book, these acts of personal choice have melded into a movement which seeks to mandate those choices for the community at large. Given that only the rich can afford them, then any such mandate would be devastating to the poor and would work to perpetuate their poverty. Contemporary "green consumerism" did not arise in an historical vacuum, as there have been several centuries of antitechnology elitism in Western Culture. A cursory look at some of the practices of the past few centuries in Chapter 7 will find that our progenitors were not "ecologically correct," nor were their attitudes and treatment of other species such as to satisfy modern sensibilities. They were neither closer to nature than we are, nor did they have a greater appreciation of it.

I have argued throughout my career that we are inherently technological beings and that the biological evolution that made us human was inextricably bound with the evolution of our early technology. Technology is an integral part of our biological being; we evolved as tool users, as I have repeatedly argued, drawing on a substantial body of anthropological literature. Technology has been with us as long as we have been human, and any concept of humans without technology is meaningless. This has not prevented it from being fashionable in the 1970s for certain types of elitist thinkers such as E. F. Schumacher to define modern technology as being "alien," harmful to human beings, and to suggest that there is "evidence" that humans are about to rid themselves of it.

It is difficult to pick up a newspaper, catch the evening news on radio or TV, or go online without encountering the latest threat to our health and well-being or to the future of civilization. If we are not populating or polluting ourselves to death, long the bugaboos of choice, then we are faced with the threat of shrunken testicles because of radiation from nuclear power plants, or declining sperm count because of environmental chemicals. One observer, speaking of the "carefully baited emotional hook," argues that "rational argument alone will not carry a message to the general public; it has to travel on the back of emotion." "Pressure groups who seek to capture public notice are

skilled in providing such hooks, often to publicize opinions that infu-
riate scientists with their misinformation and disinformation" (Emsley
2001). One tactic is to claim:

> a threat to a group we are bound to be sympathetic towards, such as
> babies, breast-feeding mothers or young children. Threats to pregnant
> women or to male fertility are likely to hook the young, while risks of
> heart disease and cancer hook those who are older (Emsley 2001).

The ongoing hysteria about populating ourselves to extinction has
moderated lately as birth rates have been falling faster than death rates,
slowing population growth and offering a prospect of a future leveling
off (or even possible decline) in about 2040 to 2050 in the range of 8.5
to 9 billion people. And at least for the time being, low and often
declining commodity prices have tempered the fear of resource
exhaustion, a scare of choice in the 1970s. The latest decade or so has
seen an ever increasing array of emerging fears and a proliferation of
NGOs (Non-Governmental Organizations), "civil society" as they call
themselves, marketing their own unique brand of fear, and using it to
create their own niche for publicity, membership and fund raising.
Closely allied are the authors of books and articles frightening the mul-
titudes, and showing the faithful the pathway to health and well-being.

Agriculture and food supply have provided a rich field for fear and
fund-raising. The existence of a plethora of food faddisms has provided fertile ground for these phobias. Some of the all-purpose fear
mongers, such as Friends of the Earth and Greenpeace, have found
enough financial nutrient in the fears they have generated about genetically modified food for more than one group to feast on. Some well
established practitioners of "civil society" like Public Citizen, not
being a leading player in the antigenetically modified-food fight, have
sought to carve out their niche by leading the fight against food irradiation, the one existing technology that could very quickly make a significant contribution to reducing the pathogens in the produce and
meat we eat, thereby making our food supply even safer and our population healthier (Tauxe 2001).

The big food scares and the widely accepted belief that our food supply is "contaminated" and unhealthy have provided an umbrella under
which authors and groups are creating their own scenarios of what is
threatening us. Among other things to fear, we are told that our soils are
being washed away into the oceans, and those soils that remain are being
depleted of nutrients and are therefore not able to provide them to the

crops. This may not really matter, however, because humans have bred the nutrients out of other crops and then have been cooking out what little nutrients were left in them. Others have long believed that using synthetic fertilizer deprives our food of some mystical, non-verifiable vital essence of life. This vital life force is available only in "organically" or "bio-dynamically" grown food. To the existing terms for these vital properties has been added a term from the Hindu religion, *Prana*, as in the claim that genetically modified food lacks *Prana*. If all these scares were not enough to put humanity on the path to perdition, we periodically have warnings about the health hazards of the way we prepare food or how others prepare it for us. One week it is Chinese food, another it is Italian cooking, with fast food always available as a subject about which someone can write a racy best seller. Space only permits the smallest sampling and briefest mention of the food fears that are continually being created. While some particular fears may pass as it becomes overwhelmingly obvious that they have no basis in reality, the fear production itself seems not to have any type of business cycle and continues as a thriving, growth industry as new fears arise to replace the old ones. Even so, old phobias rarely ever completely die but largely fade away to be revived and recycled another day for another cause.

The warnings against Chinese and Italian food were reported on the network nightly news programs. All the other food fears have been in my local newspaper in the last year or so as authors have passed through town touting the latest horror stories to generate book sales. If any evidence supports the thesis, whether or not the work was submitted to some kind of peer review or what the author's credentials are, is apparently deemed irrelevant as these facts are rarely if ever mentioned. If you were under some illusion that the food you have been eating is safe, then a report saying that it is not is news. But the bad news of finding some hidden, previously unknown danger is news, or as an Associated Press editor once said, in another context, "Plane Lands in Malawi" is not news, "Plane Crashes in Malawi" is. A preliminary indication that coffee may be a carcinogen, given at a press conference at a scientific convention, made the evening news on all three networks and was a front page story in the *New York Times*. When the same researcher designed a follow-up study to test this proposition and found no evidence that coffee was carcinogenic, it ceased to be news except for an item buried deep in the *New York Times*. When an article purporting to show possible harm to the Monarch butterfly larvae was peer reviewed for two different publications and rejected and then published as a correspondence, it instantly received worldwide media coverage, taking on a life of its own complete with ongoing costumed street theater (Shelton and Sears 2001). When one of the most prestigious scientific publications published six peer-reviewed articles finding at most negligible harm, little notice was paid to them, and the street theater is almost certain to be unaffected by the evidence (Hellmich et al. 2001; Oberhauser et al. 2001; Pleasants et al.; Sears et al. 2001; Stanley-Horn et al. 2001; Zangerl et al. 2001). One newspaper that did take notice allocated as much space to those who adamantly refused to accept the failure to confirm harm, which suggests that once set in their beliefs no lack of evidence of harm will persuade them to change, particularly when their belief furthers their organization's membership and fund-raising efforts (Pollack 2001a, 2001b).

One would hope that after being so saturated with bad news, some of the media would see good news as having sufficient novelty to warrant mention. When a television network reporter, John Stossel of ABC, dared to counter the received wisdom of gloom and doom, he was pilloried and subjected to a barrage of calumny, including being called a liar, with demands made that he be fired, by leaders of supposedly reputable environmental groups. These so-called public interest groups have ties to those most likely to benefit from the food scares, the "organic" food industry. The bad news bandits not only wish to dominate the news, they actively seek to prevent any other view from being expressed. The actions on the streets and on the campuses using the organized force of mob violence to prevent differing points of view from being expressed can be understood as an extreme but consistent manifestation of this intolerance of difference. Given their absolute certainty concerning their assertions, it is amazing how frightened they are of any differing ideas, and how they fear even being exposed to those ideas as if they were a disease.

Extreme visions of imminent peril to life on planet Earth warrant extreme actions to prevent their realization. Those who awaken the slumbering beast of fear and violence cannot escape responsibility for what logically follows from their rhetoric. In the era when military strategists and Hollywood filmmakers spoke of a doomsday machine that could destroy the earth and its inhabitants, I was asked what I would say to persuade someone not to push the proverbial button that would end it for all of us. The simple answer would be to say anything that would keep him or her from doing us all in. The idea of truth in this context has no meaning, and only the most naive and immoral would adhere to any statement but that which prevented the ultimate in wanton destruction. And need I add that imposing the ultimate sanction—termination with extreme prejudice—would also be warranted if there were a reasonable possibility that the button would be pushed and that harm to others would follow if preventative action were not taken.

The term "terrorist" has been widely used for those inflamed by the antitechnology, antiglobalization rhetoric, who have burned down buildings, destroyed agricultural fields, or wreaked havoc in research labs. It has also been loosely used against those of us who favor modern science and technology. In the aftermath of the events of September 11, 2001, it is imperative that we all become more careful with our rhetoric. However terrible murder may be, we do have terms such as holocaust, genocide, and ethnic cleansing that can ratchet up the evil and thereby provide definable differences. For terrorists, we have only one word, and however wrong and evil some of us consider the destructive acts of the antitechnology, antiglobalization activists, these acts are too vastly different to be covered by the same term used to describe the events of September 11. Whatever we call these acts, it is fair to mention that they are continuing right up to this writing (Verhovek 2001).

Currently, for some, all issues of law, ethics, morality, and human rights collapse into one overarching objective: Save Mother Earth and all her inhabitants. Those who preach the apocalyptic environmental message need to recognize the actions that their proselytizing provokes. Just as they seek to save the Earth from our actions, they have an equal responsibility to attempt to prevent the violence they have engendered.

> Such dictators derive their intellectual drilling rights from the belief that ... they have selfless motives. They want nothing for themselves—only a better tomorrow for all earth dwellers. Their belief that they alone are pure of soul is what makes them so pushy (Wolcott 1993, 124).

In a strange sort of way, one minor aspect of this book is my plea for help. No, I am not destined to an early demise by my own hand, nor will I flip out and do harm to others. But I am experiencing a rising level of frustration with no end in sight. For over a quarter of a century now, in books, articles, radio talk shows, academic meetings, and about everywhere else, I have asked the same question: If modern science and technology are killing us, why are we so healthy and living so long? I have quite literally met a deafening silence, with two exceptions. At an academic meeting, I was told that the question was too complicated to be answered in that setting without specifying when and where it could be answered. A colleague in another department said that these were just numbers and refused to discuss it further, as if life expectancies and infant and child mortality statistics were not about real human beings and those who love them.

I have asked the question that dare not be answered. Far better to simply ignore the question, for there is a very good reason to act as if it had not been asked. The evidence is so overwhelming that it is easier to ignore the question than to attempt to refute it. Accepting this evidence of betterment of the human condition into public discourse on issues, places the onus on those opposed to technology and science to come forward with an alternative explanation as to how this betterment is happening, and to prove that any proposal for an alternative way of organizing human life would be superior. And simply and starkly put, they cannot do this. However much they may proclaim that they are on the cutting edge of tomorrow, most of what is being offered is clearly a return to yesterday when things were far worse for most of humanity than they are today.

Absent any other explanation as to why we live longer and better, except for technology and science, fundamentally changes the discourse about how to advance the development of science and technology, and how to make it more effectively serve the cause of human betterment. If we are in fact engaged in a process of human betterment, then doomsday does not seem as near, nor will cries of its imminent arrival fill an organization's coffers. Within the framework of the societal process of using science and technology for human betterment, even the most rabid technophiles, this author included, recognize that the process is far from perfect and there is always room for intelligent criticism and improvement. Within the framework of scientific discourse on current issues it might just be that some of the niche criticisms might have some merit. As I have often argued, it is clear that chemicals are not killing us, though that doesn't mean that a particular chemical might not be doing more harm than good than to humankind. It is far easier to identify and remedy that which is harmful in a framework that recognizes benefit than it is in the context that fosters generic beliefs about "chemicals" or "pesticides" being life threatening. Paradoxically, the outcome of such rational behavior would be the more efficient and effective use of our technology to bring even greater gains in human well-being, and the achievement of the greater safety that the critics claim to be seeking.

The recognition of the lack of absolute certainty and perfectability is inherent in the very nature of scientific inquiry. This fact has put scientists at a disadvantage as true to their calling; they offer ideas in terms of probabilities and uncertainties, making it difficult for them to counter the presumed certainties of the ideologues in the arena of public policy formulation. This lack of certainty is too often shamelessly used by the opponents of science to imply that scientists don't know the answers but the critics do. In the debate on genetically modified food, a standard ploy of its opponents is the rhetorical question, can you be absolutely certain that no harm will ever come from it? The simple answer is of course not, since we cannot be absolutely certain of anything. But unfortunately, for many listening, the pernicious seeds of doubt and fear had already been planted, however unwarranted by objective fact.

Any concession that we are living longer and healthier lives means that the process is in some sense working. If it is not working those who oppose it are under no obligation to provide a better alternative though that would be nice. But if it is working, and is in fact working at a level never before imagined, then the critics have to provide a way to either improve the process or offer a better alternative of some sort and not some vague hint of a New Age utopia. And if technology and science have played an essential role in this betterment, then the critics might just be forced to play by the rules of scientific inquiry with realistic options being equally evaluated in terms of probable benefits, possible risks, and the level of certainty and uncertainty. I do not expect to get my question about why we are living so long answered precisely, because those of whom it is asked do not dare open their advocacy to any rigorous examination and to the rules of scientific inquiry.

It is clear globalization is operative throughout the modern economy and will loom ever larger in the future. Those who claim to be against globalization nevertheless are avidly promoting their own global agenda: Seeking to ban production of and trade in various pesticides, and calling for a global moratorium on research and development in genetic modification. It would be difficult to name any group more active in trying to interfere in other people's lives around the world than those proclaiming an antiglobalization message. I argue in Chapter 2 that the alleged dichotomy between the globalizers and antiglobalizers is false, and the conflict is not between the globalizers and antiglobalizers but between differing visions of globalization and the fundamental question of who controls resources and for what purpose.

The world has been moving toward globalization for the last six hundred years, starting with intrepid Portuguese seamen in their leaky caravels as they worked their way down the African coastline. By the 19th century a small number of European countries had acquired the power to impose their vision on much of the rest of the world in an endeavor called colonialism or imperialism, both terms now rightly considered pejoratives. Today it is widely accepted that 19th Century eco(nomic)-colonialism was wrong, but as I argue in Chapter 2, we now have a new eco-colonialism, though this time it is eco(logical)-colonialism.

Mention the terms globalization and trade, and multinational corporations and international institutions such as the World Bank and the International Monetary Fund (IMF) come to mind. What if I were to ask one's view of organizations that were actively promoting policies that would lead to an increase in the death rate of some of the poorest, most vulnerable of the world's population, with particular harm to African children? What if I further indicated that one of these organizations had a better than $100 (U.S.) million budget to influence public policy and about 2.5 million stakeholders but whose leaders, a small group of predominantly white, northern European males, are chosen by an inner group numbering in the two digit range? The latter must be a corporation and no epitaph—murderers, racists—would be too extreme to describe them. But if that organization is Greenpeace, then suddenly there must be a reason for their actions—"saving the planet," that catch-all justification for all actions—and we should refrain from leaping too quickly to any conclusions. Or should we? The attempt by Greenpeace and other environmental groups to get a global ban on DDT would have led to an almost immediate increase in deaths around the world—but mainly in poor tropical countries—from malaria, with the heaviest toll being among children in Africa where it is already the number one killer. When the final global ban on a number of chemicals was to be voted on, some environmental groups backed down and acceded to an exemption for DDT for malaria control, but others remained unrepentant and still loved by the true believers.

For this book, I have chosen the globalization of wildlife and habitat to illustrate the active globalizing actions of those who claim to be against globalization. Not only does it provide a dramatic illustration of a surreptitious globalization of resources, but it also shows that many of the contemporary forms of ecological colonialism are a continuation of treaties, laws, and policies of the previous economic colonialism.

In globalizing habitat and wildlife resources, the rights of the local inhabitants to these resources are too often trampled, and lives and livelihood are lost. Some of the more enlightened environmental organizations have begun a glacially slow process of trying to use conservation as a means of benefitting the local population economically to

cover the loss from the appropriation of their resources, but the fundamental belief that these resources are to be controlled by international
bodies, not by the local inhabitants, remains largely unchallenged. We
must never forget that one of the premises of "enlightened" colonialism was the "dual mandate" to develop the resources of Africa, for
example, for the benefit of the local population and the benefit of the
world, but that the colonial authority would be the sole determining
force of how that mandate was to be carried out. In later chapters I will
pick up on this theme of the loss of local control of culture, tradition
and resources in the name of a larger global good.

What we are talking about in Chapter 2, then, is a form of neocolo-
nialism. Those who have so freely used the term over the last decades
rarely apply it to themselves or to those who exercise control over the
habitat or wildlife resources of other people, but they do apply it to those
who make use of other resources, such as minerals, and pay royalties to
do so. In the 1970s, dependency or world systems theory was much in
vogue. They spoke about a local "comprador" class who were the
advanced guard and the minions of the foreign economic colonialism. In
recent years, environmental groups sensitive to the charge of being elite
white northern Europeans, or North Americans, or of being wildlife
groups run by European royalty and hunters, have used their ample funds
to set up branches in developing countries. These local hires, like their
predecessors' economic compradors, are then put forward as spokespersons for their country in international meetings, no matter how unrepre-
sentative they may be. They become the ones who get air time or press
coverage in the Western media. And like the compradors of old, they
have a standard of living they would not otherwise have even remotely
achieved if they had to depend on employment in the local economy.

As the NGOs in the developed countries and their compradors in the
developing countries have become more powerful in seeking to impose
their will on others in the name of defending them against a variety of
evils, from modern technology to globalization, the ideological basis for
their advocacy has become increasingly evident. Though they portray
themselves as Davids battling the Goliaths, the reality is quite different.
According to a 1998 study, by 1993 there were 28,000 international
NGOs with 20,000 NGO networks, employing 19 million people with an
income of $1.1 billion. Since many of these are purely lobbying organi-
zations, their discretionary funding for the campaigns is comparable to
that of those they oppose (Peel 2001).

On the conflict over the use of genetically modified foodstuffs, for
example, with one or two rare exceptions, the NGOs opposing genetically

modified foods have had no experience in helping poor people grow food or otherwise helping them provide adequate nutrition for their families, even though some of the NGOs existing for a quarter century or more have had terms like "food" or "agriculture" or "rural" in their names and have been raising money on the premise that they are advancing the cause of rural development. As a *Financial Times* editorial (prefacing a series on NGOs) correctly states, NGOs have a right to "lobby for their arguments, like any other private sector organization." But they have neither a "veto" nor do they have any "monopoly on claims to represent civil society" (*Financial Times* 2001). In the development community, there is now the term "bringo" for "bring your own NGO" as various interests try to claim some legitimacy and the appearance of popular support for their causes (Beattie 2001). Clearly, those with the longest and most productive experience in helping the poor feed themselves better have overwhelmingly lined up in support of the use of advanced technologies such as genetically modified food. A dichotomy has emerged between those who have efficiently, effectively and productively used resources in helping the world feed itself, and those who have opposed them by offering theoretical alternatives. More important has been the use of science and technology to transform and make more productive the agricultural resources available to the world's cultivators.

In recent years, after decades of existence, NGOs that have now ventured out into modest agricultural programs might tempt cynics to suggest that these were in response to the criticism that they were making claims about Third World agriculture without having any field experience in it. Whatever their anti-establishment slogans may be, many of these NGOs receive government funding that would otherwise have gone for economic development. Currently, one quarter of the development budget of Norway goes to NGOs while the United Kingdom spends close to 16 percent of its development budget through NGOs and receives complaints when they attempt to shift some of these resources to developing country organizations. Funding projects directly through national governments is now a threat to some NGOs in rich countries. NGOs have long been used to presuming to speak on behalf of those in need but Clair Short, the UK Development Secretary, says "the days have gone" when even a very worthy organization like Oxfam "can speak on behalf of the poor" (Beattie 2001). Far less worthy organizations continue to presume to speak on behalf of the developing countries and the world's poor, much to the frustration of the elected leaders of these countries, particularly when the NGO agenda is exactly contrary to what they are seeking on such issues as trade and development. However worthy their demands for human rights or for elimination of child labor may be, to leaders of developing countries the demands sound like disguised forms for protectionism. NGOs pressuring multi-nationals to promote human rights in a country seems to some like trying to impose change from the outside where it can only properly be done from the inside (Ottaway 2001; Landsburg 2001). And some NGOs, like Amnesty International, which have been extraordinarily effective in protecting the human rights of many individuals, could jeopardize their fine work by trying to compete with other NGOs and extending their mandate too broadly (*Economist* 2001a, 19–20).

Somehow NGOs and conservation activities are "privileged" and exempt from responsibility for the consequences of their actions and too often receive the uncritical adulation of the media (Furedi 1999). Currently, there is a rising tide of legitimate concern and protest against the government of Myanmar (Burma) about the construction of an oil pipeline and the use of the military to displace the local Karen population (Solomon 1998). New York's World Conservation Society and Washington's Smithsonian Institution are "under fire" over a nature reserve project in Myanmar "that may involve wholesale village relocation." Compared to the protests over the pipeline project, there is far less publicity about the use of the military to displace Karen for the "creation of the million-hectare corridor Myinmolekat Nature Reserve south of the pipeline corridor" (Faulder 1997, 48). The institutions may be "under fire," but their actions have received far less media attention than has the construction of the pipeline. There was a lawsuit to stop a Burma-to-Thailand gas pipeline, even though measures have been taken to avoid passing through a forest (BBC 1998). No lawsuits were reported to prevent the creation of the nature reserve.

Romantics have turned to other cultures for validation of their beliefs and lifestyles and to find what they feel is lacking in their own. This quest has focused on a variety of peoples and ways of life and includes both pre-agricultural and contemporary hunters and gatherers, pre-Columbian and contemporary American Indians, and ancient and modern Tibetans and Pacific Islanders. What would appear to be benign understandings of other peoples can actually be a falsification of them. My basic argument is that to romanticize and thereby falsify perceptions of others is not benign and can have highly adverse consequences. This is particularly true when more affluent and powerful outsiders seek to impose their own fictive visions on those who are poorer and less developed. Its most immediate and direct outcome is to keep them in their poverty in the name of preserving their "traditional" culture.

Chapters 3 through 7 focus on contemporary romantic mythologies about peoples who are economically less developed. On the surface, the impression may be to interpret these mythologies in terms of simply debunking a romantic view of nature and the human condition. There are, as we have previously noted, commonalities to these mythologies about earlier times and other peoples. These commonalities are neither accidental nor are they necessarily intentional, but they do flow from widely shared beliefs about how the world should be organized. In other words, they are less about other peoples, particularly in the case of those who have largely passed from the scene, and more as cautionary tales about what is wrong with modern life and what needs to be done to rectify it. They form a critical component in the argument for alternative lifestyles and an array of other factors defined as alternative, from medicine to agriculture. Remove the rose-colored glasses and the romantic patina, and one finds an argument about the allocation and use of resources in modern life. Once, opting for the "alternative" lifestyle was just that, an option or choice that a few made. But in recent years this increasingly strident and growing minority has sought to force its "alternative" practices on the larger community by democratic means where possible and by force if necessary. The argument about how to use resources comes in many guises and is the central and underlying theme of this book, as it is in modern life today.

Historical myths serve as social charters authenticating the lifestyles of peoples and societies. Being historically inaccurate is irrelevant, as their role is the formation of an ideological agenda that defines and legitimizes the present society. Consequently, a counter myth is not simply a misinterpretation of the past or present but a counterclaim as to how to organize society and use resources. In the discussion of eco-colonialism I argued that mythologies about wildlife habitat and the humans that share this habitat have been used to legitimize external control, a control that is no different from that which has long been denounced as imperialism. Ethnic groups so caught up in the wildlife conservation nexus are no longer allowed to define what is "traditional" for them, as others presume not only to define it but also to impose their concept of what is traditional, and thereby permissible, upon them. The further a myth differs from reality, the more authoritarian and repressive must be the means to maintain it.

A very serious question has arisen as to whether those seeking to preserve wildlife favor the interests of animals over that of humans. As I have argued in many places, the lives of children and adults who suffer and die from malaria seem not to have been given equal consideration to the concern that the use of DDT to control the disease vector could harm wildlife. To Western elitists, the fate of one Indian elephant that was captured, chained and transported "30 miles to Mudumalai Wildlife Camp in Tamil Nadu" was of greater concern than the fact that he "trampled to death 12 settlers, plundered crops and destroyed villages." The possible mistreatment of the elephant led to an "international effort spearheaded by a Hollywood actress and more than two dozen U.S. congressmen to save" him though it was less than clear what it was they would be saving him from (Schmetzer 1999). Similarly, in 1988, an icebreaker was sent by the Soviet Union, the National guard was called out by President Reagan and over a million dollars was spent by the United States government to try to save three whales trapped in the ice off the coast of Alaska. Some "wondered why the guard was called out to free whales but not hunters lost on the ice the year before (who died)" (Krech 1999a, 223; see also Kalland 1994, 181). A colleague in anthropology tells of a similar tale in Madagascar in which all-terrain vehicles were sent to a wildlife preserve, Ranomafana National Park, for an injured lemur but were unavailable to transport village children with life threatening illnesses to a hospital (Harper 2002; see also Stille 2002).

Wildlife conservation is only one of several examples that could be used to illustrate eco-imperialism and eco-interventionism. The interventions are just as blatant to prevent the construction of dams for hydroelectric power, water storage and irrigation for agriculture. Activists in distant developed lands who couldn't find the map location for the Epupa, the Bujagali or the Narmada Dams even if they were given the coordinates, will nevertheless argue that Namibia, Uganda or India do not need the power from those dams and will actively lobby against them with donor and/or lender organizations (Denny 2001; Lacey 2001). It is recognized within the professional development community that "foreign aid" does not have a domestic constituency with a vested interest in policy outcomes in the same way that agriculture or commerce has. Consequently, very small but vocal, ideologically committed activists can influence aid policy vastly out of proportion to their numbers or the logic of their arguments.

Successive chapters of this book deal with mythologies about differing peoples, and each is used to explore a particular variant of the prevailing belief system in developed countries that gives rise to the mythologies and the differing ways that they are used to promote a particular form of resource use and control over economic activity.

Vegetarianism is seen by many as the cutting edge of the "natural" lifestyle of healthy, environmentally friendly consumption. It is also seen as the way to more effectively use the world's limited land resources to feed a larger population, and therefore it is yet another argument as to how the world's resources should be used. I will challenge these assumptions. Vegetarianism is also considered to be natural because many wish to believe that our hominid ancestors were, and our nearest primate kin are, vegetarians. It was once thought that *Homo sapiens* were originally herbivores and became omnivorous by some accident of evolution. To the committed vegetarian, evolution made a wrong and unnecessary turn at that juncture. In Chapter 6 I examine an emerging body of evidence that argues that our closest relatives (or one of the closest, along with bonobos and gorillas), chimpanzees, are hunters of monkeys and hence are meat eaters (DeGregori 2001). Now that vegetarianism has taken on the mantle of a larger "green" ideology, instructors of courses in primate behavior regularly have students who very strongly object to any claim of other primates being meat eaters, no matter how powerful the evidence may be. Further, just as it is now accepted that tool using was a necessary part of the evolution and the emergence of *Homo sapiens*, it is now being argued that meat eating was essential for the emergence of modern humans (77–81). I will simply argue that vegetarianism is a matter of choice for those who wish to practice it, but it is neither "natural" nor is it a necessary condition of feeding the world's population, either in the present or in the future.

The quest for the natural has its corollary in the pursuit of the authentic. Once again, we have a lifestyle that is restricted to an elite group and is not sustainable if applied to or pursued by the larger community. The pursuit of authenticity in what is presumed to be the simple life—"live simply, so others may simply live"—can actually be quite expensive and beyond the reach of the larger community. A car with bumper stickers the read "Visualize Industrial Collapse" and "Question Technology" is interesting because of the contradictions the slogans embody. No industry, no car, and "question technology" is undoubtedly rhetorical, as the questioners have shown that they already have the answer.

Many seek authenticity in other cultures, and if the members of these cultures wish to experience change they often find their "traditional cultures" defined and imposed upon them. Political and economic imperialism has been superseded by "green imperialism" and what might be called the "imperialism of authenticity," which I examine in Chapter 7 as yet another case of defining how resources should be utilized and who decides on the form of their use.

The other side of the coin of elitism and authenticity is snobbery and disdain for the popular, which is deemed to be vulgar in both senses of that term. In opposition to the elitism and snobbery of the "primitive" is the promise of modern technology that offers the possibility of both excellence and abundance for all. In the concluding chapter (Chapter 9), I challenge the notions of elitism arguing that much that is popular can also be very good, and offer a vigorous defense of modernity. The promise of technology goes beyond abundance of what is good, but it also intensifies and extends our senses and allows us to deepen our understanding of ourselves and the universe around us. This is contrary to the thesis of technology being alienating, which has become the coin of the realm in some circles.

CHAPTER 1

Green Consumerism

I cannot give any scientist of any age better advice than this: the intensity of a conviction that a hypothesis is true has no bearing over whether it is true or not. P. B. Medawar, Advice to a Young Scientist.

(Whelan 1993, 67)

Elitism and Being "Socially Responsible"

"Socially responsible" has become the catch-all phrase. There are socially responsible money market funds (including those providing for religious "stewardship"), socially responsible collecting, environmentally correct shopping, eco-gifts, and a long distance phone service that funds rainforest protection (*Buzzworm* 1990, 9, 17, 21; Barrett 1990, 4–5; *Aperture* 1990, 80; Scherreik 1998). There is socially responsible international "ecotourism" that severely restricts how much its participants can spend in the local economy, particularly for craft items. In the name of protecting the indigenous culture, the population is deprived of many of the economic benefits of tourism. People can stay in "ecofriendly" resorts or hotels (Abreu 1994).

There is also minimal impact camping with differing philosophies as to the disposal of human excrement. Some haul it all out even where it could be recycled through the ecosystem. Some allow for burial of excrement at appropriate depth but not at elevations above the tree line, where it must be smeared on the north wall of rock faces so it can be degraded by sunlight. Not only are these "responsible" forms of consumption expensive, they are time consuming and difficult due to all the consumption boycotts and taboos (Milbank 1991, 1, A9).

3

Hotels are not always as ecofriendly as they claim to be:

The sewage material from one Ngorongoro hotel, for example, is dumped at a "safe" distance from the tourist hotel and allowed to flow onto neighboring grazing grounds and Masai settlement areas. In other parks, sewage material from campsites is simply thrown into the river from which wildlife, livestock and local communities draw their drinking water (Kamuaro 1996, 62).

Ethnic groups on the margins of Ngorongoro and many of Tanzania's wildlife preserves are experiencing a decline in their livelihood and numbers (Odhiambo 1999; Brockington and Homewood 1999).

Most ecotourists are unaware of the tampering with the "natural" environment that is often necessary so that the tourist facilities are safe for our nature lovers. (The term natural is in quotation marks since most of the world's environments have been anthropogenically transformed one way or another.) Many lodges with nature trails, etc., have regular patrols to find and kill poisonous snakes so that the visitors can safely enjoy the "pristine unspoiled" beauty of nature. I was recently at a meeting at a lodge in a wildlife park where about a dozen heads of state were in attendance. The military and police security were very much in evidence. Unseen and unknown to most but almost as important was an outer perimeter of patrols seeking out and killing poisonous snakes.

Ironically, ecotourism and the quest for authenticity can create greater honesty in tourist promotion brochures. A tourist agency that originally promoted Caribbean vacations as a bit of paradise—"sun-drenched beaches, shady palms, luxury hotels and loving couples"— now describes them for the back-to-nature lovers as follows:

Ants, mosquitoes and cockroaches thrive in hot climates and while usually harmless are sometimes a nuisance. In the long hours of sunshine, lack of rain can mean erratic water and electricity supplies—really hot water is rare (Mowforth and Munt 1998, 57).

These "shortcomings" are to be seen "as a challenge and as an enriching experience rather than as a reason for complaint" (Mowforth and Munt 1998, 57). Probably, the tour company is now exaggerating in the other direction as most Caribbean tourist destinations have modern facilities to cater to the tourist needs. However, to go to far away, difficult-to-reach places where few others visit and where there are few of the comforts of modern technology carries prestige even if one has to fake it.

For those who are socially conscious but want something other than ecotourism, they can try the "virtuous vacation" where they will pay to go off and work by improving migrant housing or teaching English in another country. Unlike ecotourists, most of these endeavors, which seem to mimic the work of the Peace Corps or Habitat for Humanity, involve doing traditional tasks such as construction or teaching. Nothing New Age here! In some instances, those unskilled in the crafts they are to perform may turn out to be more of a burden than a benefit to those they seek to help (Carey 2001). But no one can reasonably fault the effort or intent, and the solution would seem to be better pre-vacation training by those organizing these efforts.

What is "wild" in wildlife preserves is there because of tourism and not as a vestige of some untamed nature. "In a way, tourists contributed as much to the *creation* of 'wild' landscapes as to their preservation: Simply by expecting to see game and making their wishes known to administrators, tourists initiated a management regime that introduced more and more game to park ecosystems" (Warren 1997, 143). Warren is speaking of wildlife preserves in the United States, but his observation has near universal validity.

> Ecotourism planners recently put local participation in decision making high on their agendas, this has been mostly done to confuse dissent. Rarely have local people been involved in planning and implementation of ecotourism ventures (Kamuaro 1996, 62).

In many respects, ecotourism is an oxymoron. Creating and maintaining the facilities for ecotourism require the revenue of a large volume of tourism that places strains on the environment. Ecotourism neither protects the environment, as its proponents claim, nor does it help local cultures. "Ethnic groups are increasingly being seen as a major asset and 'exotic' backdrop to natural scenery and wildlife." The basic purpose of ecotourism is to maintain local cultures "as archaeological artifacts, stimulating the tourist's nostalgic desire for the authentic, the untouched, the primitive and the savage." It engenders the "false notion" that the local peoples are "willing and available for 'discovery' by tourists" (Kamuaro, 1996, 63).

Though ecotourism, if done properly, may have short-term economic benefits, it is highly questionable whether or not it aids development in the long run. One observer has suggested that the "promotion of ecotourism in the developing world represents a 'degraded' kind of development, which ties people to their natural environment and offers no way to take those societies forward." For the past two centuries, the West has developed "industries and societies for human benefit" but now, countries in the South are being "deprived of the ability to choose to follow a similar course." Serious questions need to be raised as to "whose interests are served by policies that prioritize sustainability and conservation in development."

> The emphasis on degraded forms of development seems to reflect the desire of Western governments and people to preserve areas rich in bio-diversity for their own benefits, while they themselves live with the benefits of development (Craig 2001).

These polices might "save some charismatic species from extinction or beautiful forest from destruction, but it will do little to fulfill the aspirations of people in the South" (Craig 2001).

Back-to-Nature Urban Style

For after-hours entertainment when nature's noblemen or women return to New York City, there are places like the Wetlands in Manhattan which was described in a blurb in the *New Yorker* as follows:

> Two floors of no-nuke veggie entertainment and consciousness-raising. Listen to bands play in the Summer of Love-muraled back room; buy a tie-dyed T-shirt at the Volkswagen-bus curio shop; catch up on current events at the bulletin board and community calendar. Or, if you'd rather just be mellow, crash out in the basement hippie love pad (*New Yorker* 1990a, 10).

Presumably, your cause or mine is the new mating call for our urban nature lovers. The Vegan Society has given its seal of approval to cruelty-free "vegan condoms" that are made from cocoa powder instead of latex for those desiring a politically correct sexual encounter. No animal ingredients or derivatives (including casein—a milk protein—which is used in the processing of latex) are used to make vegan condoms (BBC 1999c). At the Hard Rock Casino in Las Vegas, Nevada, there are "morally responsible slot machines" for New Agers with a yen for gambling. The slot machine's proceeds (or at least some of them) go to save rainforests (ABC 1995). For the truly avant garde in California, there was an "evening of magic to benefit the videotaping of a dolphin-assisted human birth" complete with a "cosmic concert", a "laser lite show", and a "silent auction" (*New Yorker* 1990b, 37). For the depressed New Agers, there are "dolphin therapists" (Kalland 1994, 172).

For your chic friends, there is a "recycling fairy" birthday card printed on recycled paper using "soy-based inks" (*Earth Care Paper Inc.* 1990, 20, 26). There are scientific studies, however, that argue that incinerating paper products for their energy content is better for the environment than recycling them (Pearce 1997; see also Scarlett 1991). The New Age Shaman can buy a portable Sweat Lodge "that you can take anywhere" (Root 1996, 91). Among the many types offered for sale is the Sweat Stone Lodge:

> Comes with FREE carrying case and handbook on the history and health benefits of Native American Sweat Lodge Ceremonies (91).

The above can be shipped immediately and comes with a "30-day money-back guarantee." There is an 800 number and they take MasterCard and Visa (91). Retreating to a tepee in one's backyard as a means to relieve tension is the latest fad (Gray 1997). There are also advertisements for "Hawaiian Shaman Training" for those of you who are "ready to develop your powers of mind and body." Presumably it helps if your heart is "open to the magic and mystery of nature" (Root 1996, 96). In California, one can buy a "*chi* machine with a vibrational frequency that will clean your *chakras* in only ten minutes" (Parkes 2000).

We can no longer sit down and enjoy a meal without being concerned about which political statement we are making by our gustatory actions (O'Neill 1990). Nor can we enjoy a holiday meal without being warned of the dangerous "chemicals" that we are ingesting in an article appropriately titled "The Grinch that Stole Christmas Dinner" (Radford 1997). None of these endeavors even remotely considers lower income families and their needs.

Consuming to Save the Environment?

Ours is a civilization of the machine. This is true not only for the developed countries but increasingly for countries all around the globe. For Western countries, this transformation has been taking place since the Middle Ages. It was the Renaissance man who "invented and brought to near perfection the civilization of the machine" (Oates 1973, 38). Joyce Carol Oates adds:

In doing this, he was simply acting out the conscious and unconscious demand of his time—the demand that man (whether man-in-the-world or man supposedly superior to worldly interests) master everything about him, including his own private nature, his own "ego," redefining himself in terms of a conqueror whose territory should be as vast as his own desire to conquer (38).

Many of the antitechnology, return-to-nature partisans of contemporary times are not mechanics; they are affluent. In their interaction with nature they are not averse to using the latest and most expensive high-tech materials. Alston Chase describes gear that modern wilderness warriors take with them to the mountains:

Dacron fiber-filled sleeping bags; rainwear made of Gore-tex (developed from Dupont Teflon frying-pan coating) or Bion II (from artificial-heart research); lightweight tents made of extruded aluminum, polypropylene, Mylar, polyester, or rip-stop nylon, and coming in every conceivable shape—geodesic domes, A-frames, pyramids, tunnels, cylinders, hypars; fishing rods of graphite fiber (developed for jet fighters); chemically processed, freeze-dried food; fiberglass cross-country skis; backpacks with internal or external stays made of carbon fiber (from NASA research); reflective Texolite (from space capsules); gold-plated Sierra drinking cups and night lamps, and so on endlessly (Chase 1986, 330).

Our modern explorers of nature can also take any of a variety of expensive mountain bikes, including one that folds and fits under an airplane seat. Or they can take canoes made of "preimpregnated Kevlar" with "highly sophisticated hull shapes." They can wear boots made of "synthetic materials that keep out water while still providing ventilation" (Baldwin 1985, 57–63). Contrary to the New Age rhetoric of building a "transindustrial paradigm" and "connecting our social, spiritual, and ecological visions," by those Chase calls the New Pantheists or the California Cosmologists, the frontier research on metallurgy, industrial chemistry, engineering, and other sciences is an essential ingredient in their life in nature (Chase 1987, 297).

To acquire the "natural" or "real," be it in construction with expensive stone or wood, or in foods, eating only the rare or organically grown—these natural lifestyles are expensive because the means for providing them are extremely limited, making it a way of life possible only for a very small portion of the world's population. The irony is that some of the practitioners of this lifestyle in the United States call it "voluntary simplicity" and are under the illusion that a variant of it is a prescription for the solution to world poverty (Frieden 1979, 181–83). *Time* magazine had a cover story on "The Simple Life." A perceptive correspondent for the *New Yorker* made an "unofficial tally of *Time's* 'expensive, high tech and sophisticated' stuff, as against the new simplicity's 'recyclable, cheap, plain and nostalgic' stuff." The results were:

'Recyclable, cheap, plain and nostalgic' goods ... : $459.40.

'Expensive, high tech and sophisticated' equivalents: $145.83.

He concluded that he didn't think he could "afford the simple life" (the *New Yorker* 1991b, 30; see also Carlson 2000).

Criticism of modern technology in industrial countries is an affordable luxury for the affluent who rarely find themselves foregoing the benefits of modern science and technology. Among the leading advocates of the simple lifestyle have been the very rich (see for example Rockefeller 1976, 61–65). There was even a book titled *Voluntary Simplicity* that praised the virtues of this spartan existence (Elgin 1981). Praise for it on the cover of the book *Voluntary Simplicity* included a couple in which the male later received a divorce settlement of over a million dollars (some estimates are much higher), which should keep him in simplicity. She was later remarried to a billionaire, so both should easily be able to afford simplicity. "By its very nature then, voluntary simplicity has been and remains an ethic professed and practiced primarily by those free to choose their standard of living" (Shi 1985, 7). "Consumerism with a cause" is a practice for the "well-heeled and the committed" (Dermansky 1991, 64). Unfortunately, the citizens of poorer countries do not have the means to opt for technologies that are aesthetically pleasing to the affluent.

For some, affluence carries guilt. The New Age/Back to Nature rhetoric allows its practitioners to believe their personal enjoyment of wealth, such as a $1.2 million Italian Renaissance house, is a spiritual act as part of a service to humankind since the house will be used for "benefits for causes such as protecting the environment and combating homelessness" (Schwadel 1989; see also Slesin 1989). Another "healthy house, that's high in style but low in chemicals" costs 35 percent more per square foot to build 6,000 square feet on an island (Brown 1990). Earth-friendly houses can be made with "natural building materials," covered with "house paint made with powdered milk," and equipped with "earth conscious products," such as solar refrigerators and composting toilets (Holt 1998; Fletcher 1998).

Building large (12 to 18 thousand square feet) ecofriendly homes— "Muscle houses trying to live lean"—with all the amenities, has become quite the fashion in places like Colorado where people have "great respect for the environment." As Green has become "mainstream, more homeowners want to participate but without sacrificing their amenities." "These jolly green giants are live-in contradictions, touting the latest energy-efficient accessories like photovoltaic roof tiles while admitting indulgences like climate-controlled wine cellars and motorcade-size garages." This isn't for those who are still emotionally part of the Woodstock generation, and "green" is no longer just for hippies. Today's "big green houses are far sleeker, with both streamlined photovoltaic roof tiles for electricity and solar water heaters integrated into the architecture" (Iovine 2001).

For one proud owner, the ecofriendly features may have boosted the cost of the house by 20 percent, but it was worth it. For him, "it's like a little star in my moral conscience." Even the normal human function of disposing waste can become a moral statement. "When I flush the toilet or send scraps down the composter, I feel like I am doing something important" (Holt 1998). Now that is a life that we can all envy: having the ability to make a statement about the environment and "doing something important" whenever one sits down on one's own commode to eliminate bodily waste. I am jealous! In addition to ecofriendly construction, we have "healthy and environmentally sound" renovation (Lyman 2000).

The inhabitants of these modest abodes can get the latest fashions such as a $175 signature scarf from a luxury manufacturer who raises money to save the rainforests, and whose products are allegedly environmentally benign (Hochswender 1990). For everyday shopping, there are now "Restoration" stores where the items for sale are arranged not in terms of functional categories but in terms of the lifestyle that the consumers wish to create for themselves (Brooks

1999; Cassidy 1999). In India, the elites go in their chauffeur-driven cars to small shops to acquire the latest in homespun garments, "ethnic chic" (Crossette 1989). For Americans, there is "Ecosport, makers of the world's first line of organic unmentionables."

> Besides being allergy-free, the drawers are made of cotton grown without chemical pesticides and fertilizers, and are either unbleached or colored with only vegetable dyes.

> They come in reusable cotton sacks, which the company says can be used to hold marbles and art supplies (Thomas 1993; see also, IATP 1995).

These vegetarian unmentionables would certainly be attractive to any self-respecting, ecologically conscious Lothario committed to only environmentally benign conquests. "Frankenpants," unmentionables made from genetically modified cotton, would put the wearer beyond the pale of any further interest to those who care about the environment and the future of humankind. The socially conscious can buy eco-aware cosmetics and any number of other expensive eco-correct products at chain stores, such as The Body Shop, which have prospered in catering to the socially conscious affluent. One can also buy "bird friendly" coffee that is coffee "grown on tree-shaded coffee plantations which provide habitat for birds" (IATP 1997c). Shaded coffee trees have a greater likelihood of fungal infection owing to the increased humidity which is a problem if one also wants the coffee to be "organic." However worthy this effort to promote bird friendly coffee—and it does appear to be worthy—it should be noted that the head of the non-profit NGO promoting it also heads a for-profit enterprise which markets the coffee. Unfortunately, this is not the only instance where those in leadership positions in non-profit so-called "public interest" groups are involved as consultants to, or as owners or executives in, organizations that profit from the policies that the NGO (or coalition of NGOs) is promoting. This has been particularly blatant in the various food scares over pesticides or genetically modified foods where those frightening us also have financial ties to the organic food industry. When a product of the modern science and technology they oppose, such as genetically modified food, demonstrates a clear superiority over traditional products, they either find ways and reasons to oppose it or remain deafeningly silent on its benefits.

When a variety of transgenic rice with Vitamin A is developed with a potential to benefit tens of millions of Asian children, one revered antitechnology pundit first argued that it provided too much Vitamin A and then later argued that it did not provide enough, as if even a modest addition would not help in protecting children. Groups allegedly concerned about toxins in our food have simply failed to mention that the transgenic corn that they oppose is vastly less infected (30 to 40 times lower) with *Fusarium* ear rot, a fungal infection that produces toxins, called fumonisins, which are often fatal to pigs and horses and can cause esophageal cancer in humans (DeGregori 2001). Groups which purport to be "public interest" (and have that tax status) owe it to the public to present all sides of an issue, particularly one involving safety, even where they are actively involved in a larger debate on the issue. It may seem unfair to some readers to suggest that some of these organizations are more concerned about the membership and fundraising potential of food safety scares than about providing vital information to the public who must make the decisions, both as consumers and as citizens in a democracy.

An alarm has been raised about the environmental dangers of Chanel No. 5, "the perfume that inflamed a million male fantasies after Marilyn Monroe said it was the only thing that she wore in bed" (Bell 1997). This "most French of French perfumes," which is also the best selling fragrance in the world, is being attacked by a group in Paris group named "Robin des Bois" (translated as "Robin Hood" in the wire story) which has "accused its makers of endangering the Brazilian rainforest" (Bell 1997). They are "threatening to launch an international boycott of the famous scent" if the manufacturers do not agree to use "a synthetic substitute" instead of the essential oil derived from a rare "tree 'being placed at risk by the greed of the perfume industry.'" It is not at all clear why some groups win high praise for helping to preserve the tropical rainforests and its indigenous inhabitants by using materials from the rainforest in their products while others are condemned for doing the same thing.

The ultimate, quite literally the ultimate, in green consumption is an "Eco-Coffin" made from "recycled fiberboard—no tree has to die just because you do" (Heazle 1997, 84). Quite the contrary, with a biodegradable coffin and burial in the woods, a person's remains provide nutrients for trees.

> You can assemble your final resting place yourself, using starch based glues, instead of the toxic, synthetic variety. This enables the coffin to decompose within weeks of burial (84).

The non-toxic starch-based glue means that we can be buried without fear of being threatened by carcinogens. Even more ecofriendly is a coffin "made of chicken excrement for garden interment" (Madison 1997, 84). There is also a Web site where one can "check out the Natural Death Handbook on the Internet" (85). In many respects, this form of "green" consumption makes sense for those whose personal or religious beliefs do not require a particular form of interment. (However, a caveat or two is in order here. The coffin and burial may violate local ordinances that exist for very good public health reasons. Further, if the coffin is buried deeply, the nutrients may be more likely to contaminate the groundwater before the roots of plants can grow down to harvest them. If the burial is too shallow, the plants should be able to use the nutrients, but there is the danger of disease or that a subsequent gardener or dog might inadvertently dig up a partially decomposed Uncle Harry.)

A Swedish scientist has come to the rescue of those who wish to promote sustainability with an "environmentally friendly form of burial" that has been "approved by the Church of Sweden."

> In the new green method, the body is immersed in a bath of liquid nitrogen, producing up to 65 pounds of pure organic matter, which is put into a thin, easily degradable coffin. This is then buried near the ground surface and enriches the soil in the same way as autumn leaves (Reuters 2001; CBC 2001).

Other than the fact that the human body is already organic matter, it would probably be correct to say the it is now in a form more readily available to be recycled by being taken in as nutrients to promote plant growth. In conventional coffin burial, "the body takes between 50 and 60 years to decompose" while it is alleged that "cremation emits poisonous gases with unknown effects, making it even less eco-friendly" than current practices (Reuters 2001; CBC 2001).

There are many other forms of "green" consumption that make good sense, such as unbleached toilet paper or mulching grass clippings rather than bagging and trashing them. Such practices save money and the environment. It is unfortunate that more emphasis is too often placed on expensive elitist "green" consumption and not on practical, cost-effective options.

Consuming to Save the Rainforests?

To the believers in "green" consumption, ecology stores are not only saving the environment but they can return us to a perfect world of our youth that we have lost. There is some debate even among environmentalists as to whether "marketing the rainforest" is really an effective way of protecting it (Dove 1994; Clay 1992). While some environmentalists promote the use of products from the rainforest, others protest their use. Thus, the organization Cultural Survival believes that "market-oriented strategies" can be used to "protect endangered peoples and habitats at the same time," while Survival International argues that these marketing strategies are "at best a money making gimmick and at worst a harmful idea which ... could lead to more destruction" (Dove 1994, 3).

The claims for "natural" products in food and cosmetics include the claim that they help preserve endangered environments and peoples by buying rain forest products and that they are humane because they do not test their products on animals and are "ethically sourced" and "cruelty free." Both of these claims are open to serious question. The producer of the product may not test their products on animals, but they use ingredients that were tested on animals, as the product could not have been certified for sale as safe if the tests had not been conducted.

Saving the rain forest and its inhabitants (or other environments and peoples) is not always benign. In the best of circumstances, only a very small percentage of the price of the final product goes to the rainforest providers. Often these providers are commercial ventures, sometimes even multinationals, and run by other than tribal people. There are rarely any contracts, and there are instances where tribes have been left with a large stock of unsellable product. The situation is bad enough that the results for the tribes have been deemed to be disastrous by an anthropologist who has studied Amazonian Indians and by Survival International. The "broken promises" have led to lawsuits by tribes and their leaders against "socially responsible" businesses (Gamini 1997; Stackhouse 1996; Durham and Rocha 1996; Petean 1996; Survival International 1994; Entine 1996b, 31–35; Entine 1995, 47).

To one author, "whenever a forest product becomes valuable in international markets, elites are likely to appropriate it and leave only products of little value to forest dwellers. Marketing rainforest products ... perpetuates the process of leaving to the forest dwellers the resources of least interest to the broader society." Dove goes on to label "green shopping" as a "dangerous distraction from the political and economic changes that must be made to encourage conservation of the world's tropical forests and improve the lot of the people there" (Dove 1994, 1).

Referring specifically to Ben & Jerry's "ill-conceived 'rainforest harvest'," Jon Entine argues that "reckless idealism can lead to far more significant disasters." Their nuts ended up being bought from commercial sources, including "some of the most notorious, antilabor agribusinesses in Latin America," one of which was "convicted of killing labor organizers" (Entine 1996a). The price for the nuts fell as commercial interests "flooded the market ... cutting the income of native tribes who did harvest the nuts," forcing them to sell more land rights to make up for the shortfall. The Ben & Jerry's ice cream company, in touting its "World's Best Vanilla" in "eco-pints," has campaigned against dioxin in foods, arguing that there are no safe levels of toxins in our foods (B&J 1999). Two mischievous researchers bought their "World's Best Vanilla," had it tested at an independent laboratory where it was found to have dioxin at a level 190 times the daily intake deemed a "virtually safe dose" by the EPA (Gough and Milloy 1999). "Diluted ice cream" from cleaning Ben & Jerry's machines was not washed down the drain; it was given to local farmers to feed to pigs. "Piglets that happily slurped Ben & Jerry's Homemade sugar water never made it to 600 pound adulthood," dying at 200 pounds of arteriosclerosis. The pigs that were slaughtered "yielded a fattier pork" as would be expected (Entine 1996a). This is another instance of trying to be a do-gooder without first testing the consequences of one's attempts to do good.

Socially Conscious Consumption

The irony of the use of homespun cotton as a symbol of the natural and simple in the United States and India is that the growth of cotton is one of the largest users of pesticides and it can be argued that over the lifetime of a garment, from its creation to demise, those made of synthetic fibers involve less use of energy than do garments of "natural" fibers. And wouldn't you know it, when a genetically engineered cotton is developed that greatly reduces the use of pesticides, the same environmentalists militantly oppose it and call its products "Frankenpants." "Organic" cotton was selling at $1.30 per pound in late 1997, while regular cotton was selling at $.74 per pound (For "organic" cotton and the problems of growing it, see Bose 1994 and Pleydell-Bouverie 1994. For the 1997 prices of cotton, see IATP 1997c). At a cost of $2 extra per kilogram, there is also "predator-friendly" wool obtained from "ranchers who certified that they had not killed any predators in order to protect the sheep" (*New Scientist* 1995, 13). One wonders what these groups would require for the labeling of "virgin" wool.

For those who wish to have their pets participate in saving the rain-forest, there is a "snack for dogs with nuts from the rain forest" called

Bowser Brittle made by Dandy Doggie (*New York Times* 1990; *AP* 1990). There is a chain of bakeries catering to dogs, Three Dog Bakery, with many locations. Another North Carolina canine bakery, Barbara's Canine Catering and Dog Bakery, welcomes dogs with "all natural cakes, biscuits, minipizzas and bottled water," though "toilet water" is offered as an option. "Treats range in price, with a dozen minipizzas for $6.50, a 10-ounce bag of dog biscuits for $4.00 or a dozen carob drizzled peanut butter snacks for $5.00." At $6.24 a pound, the dog biscuits are more expensive than most cuts of meat, though the dog biscuits are more politically correct. No self-respecting Yuppie would wish to have a politically incorrect dog that preferred steak to an all natural product. These treats can be ordered via the Internet from the firm's Web site, from the firm's catalog, from a natural products catalog, from supermarkets in the region, or customers can have them specially baked and catered for a party (Postman 1997). Not to be outdone, Houston, Texas had "The Bone Appetit Bakery" which announced a "'Howl'aween" pet party with contests and prizes for pet tricks and costumes for Halloween 1998.

Though there is no evidence of dogs or other pets being harmed by "chemicals" in their food, tragically, at least 26 dogs have died from aflatoxins—the very natural and dangerous toxins produced by fungus that we discuss throughout this book—in their dog food (AP 1998). The actual number is probably much higher since there are no legal requirements for autopsies to be performed on pets because of sudden unexpected deaths. Earlier in 1998, there were concerns about aflatoxin contamination of fodder for horses because of the fungal growth on feed grains, resulting from plants stressed by extreme weather conditions.

For the New Age pet owners, there are "holistic veterinarians" and even a how-to book of homeopathic first aid for animals. For one's cat, there are "wheat-based" litters that are "scoopable, biodegradable and flushable" as well as "other dust-free, biodegradable alternatives." There are several "natural" pet care products including "herbal shampoos, collars, sachets and sprays using citronella, eucalyptus, pennyroyal, tea tree oil and other aromatic oils to repel fleas and ticks." For those who cannot find these products at their local pet store, they can consult Whiskers, "a catalog of holistic products for pets" (Rembert 1998; Walker 1998).

A variety of herbs are available for our pets, as our pet stores have come to resemble pharmacies (Siebert 2000). With most of these herbs, there are few clinical studies on their benefit or harm to humans and even fewer studies of their impact upon household pets. We do know that some common products like aspirin, helpful to humans, can be deadly to other animals. Simply stated, the New Agers who use these products and provide them to their pets are engaged in a grand experiment, the outcome of which is far from certain, yet they wish to invoke the precautionary principle against products of modern science and technology about whose safety we have far more knowledge.

New Age, Nature and Consumption

The alleged New Age love of nature is conceit that often conceals gross ignorance. Dr. Helen B. Hiscoe argues that "the underlying assumption of those who value things because they are natural is their belief that nature is a benign force whereas human power tends to be evil. They equate naturalness with goodness" (Whelan 1993, 63).

One of the latest forms of New Age nature consumption is the purchase of high-priced tapes and compact discs of the sounds of nature. "Nature recordings, once a New Age cottage industry, have hit it big. Americans can't seem to get enough, spending more than $100 million a year for recordings of whales, waterfalls, howler monkeys and thunderstorms." They are often labeled as to the location of the forests or seashore or waterfall that is presumably being recorded. However, nature does not often cooperate and can be boring if not appropriately edited or assisted. "Unfortunately for purists, many of the biggest sellers are concocted more in the studio than in the woods" using a stereo recording of a "toilet flushing" because "it sounded more like a stream than the streams did." Other compact discs were created using a digital database to create a "majestic, thundering surf" or to create a 30-minute thunderstorm from an 87-second roll of thunder. Those who actually record "nature" use expensive high-tech equipment such as a "$7,000 binaural microphone" (Ortega 1995, 1, A6).

Saving the Environment

The new elite lovers of nature can always rationalize their consumption, however wrong they may be, while condemning the livelihood and consumption of others. They do not trust the intelligence of their followers to think and act correctly without the guidance of their supreme leaders. One couple, known for their fearful prophesies about population growth and global famine over the last three decades, offer their readers a list of "take home messages" and sample letters to your congressman (Ehrlich and Ehrlich 1990, 237–245). There is an organization that, for a fee, provides monthly "action alert" postcards for those who "don't have time to research the issues." With these outlines, one is supposed to write one's representatives telling them what others have told you to think. Or, what if you "don't have time to write? You can still make a statement. Earth-Cards (Conari Press, $6.95) contain pre-written and pre-addressed postcards urging legislators and officials to take action" (Javna 1990). The question is, whose statement are you making, yours or someone else's? Of course, you are putting your name to it.

One author who promotes the rights of alligators and pristine rivers over that of humans justifies his fax machine "on the premise that it makes for graceful, environmentally sound communication—an advanced way to do with less" (McKibben 1989, 190). He and his wife decided to forego their planned "wood-fired hot tub" as their contribution to global salvation (188). Those of us who previously had no intention of having a wood-fired hot tub can now willingly offer to make the same sacrifice. McKibben was blissfully unaware of the chemicals that fit his conception of being toxic (including ozone-depleting CFCs) and polluting energy involved in the production and utilization of fax machines at the time of his writing (For criticism of McKibben, see Easterbrook 1989; Cowley 1989; DeGregori 1989).

> For if you have embraced a creed which appears to be free from the ordinary dirtiness of politics—a creed from which you yourself cannot expect to draw any material advantage—surely that proves that you are right? And the more that you are in the right, the more natural that everyone should be bullied into thinking likewise (Orwell, *Polemic* 1947, 16).

CHAPTER 2

Racism, Elitism, and Environmentalism

Some have assumed the right to speak for those who presumably cannot speak for themselves. The experience of an author exploring the plight of the Huaorani of the Ecuadorian Amazon, is instructive. He speaks of the "many environmental and human-rights" NGOs who claim to be acting in support of the Huaorani and their land.

> Letter-writing campaigns, boycotts, lawsuits, grants, and foundations were being pitched and caught by the likes of CARE, Cultural Survival, The Nature Conservancy, the Nature Resources Defense Council, Wildlife Conservation International, the Sierra Club, the World Wildlife Fund and dozens of other organizations, including RAN itself (Kane 1995, 10–11).

Kane adds that, "in terms of both cost and money, the money involved was substantial—tens of millions of dollars—and the fighting bitter. In spite of all the "ruckus being raised," when Kane contacted these organizations, he could find nobody who "knew how to contact the Huaorani" nor could he find anyone who "knew what the Huaorani wanted" or "who the Huaorani were" (11).

> For American environmentalists committed to giving all creatures great and small a voice, few things make green activists more uncomfortable than charges that racism exists within their ranks (Wilkinson 1999).

Thus begins an article in the Christian Science Monitor titled "Charges of Racial Insensitivity Beset Environmentalists." The article relates "two inflammatory incidents in the past month" (November 1999) in which an executive director of one environmental group and a regional director of another made racist remarks about Hispanics. The executive director "stepped down" while the regional director "was temporarily suspended." "These comments have added to the perception among civil rights activists that the caucasian-dominated conservation movement has been slow to integrate people of color" (Wilkinson 1999).

The increasing wealth from the Industrial Revolution and its aftermath has allowed many in developed countries to rise out of poverty. This new, more widespread distribution of wealth has created the opportunity for a new kind of elitism that is more than a little tinged with racism. By this we mean that many more people have found a host of new reasons for snobbery, elitism, and an overall sense of superiority to others, including entire groups of people. Among the most notable of these new forms of snobbery and elitism is the emerging environmental elitism, which too often is translated into institutionalized racism.

Many environmental groups have tried and succeeded in becoming mass movements, some having as many as 5 million members worldwide with annual budgets well in excess of $100 million. However large these numbers may be, they reflect more the emergence of a large middle class in developed countries and increasingly in developing countries. The very size that some of these organizations have achieved in Europe and North America makes the near total absence of unskilled workers and racial and ethnic minorities all the more glaring. Environmental groups often wonder aloud why they are unable to attract significant numbers of minorities and members of the working class to their cause. The answer is in their own racist elitism, which most fail to recognize. Year after year, environmental groups admit to their previous neglect of the interests of workers and racial minorities, and claim that they have now learned their lesson and have changed. There are, of course, a vast number of environmental organizations with a wide spectrum of differing views and perspectives not all of whom are guilty of these sins.

For the conservation/environmental movements, one can be quite informed about them and even be a member and/or contributor and still not know of their support of shoot-to-kill (humans, not animals) policies and their assuming the right to define what is traditional in another culture. Some know about shoot-on-sight policies and strongly approve. Need we note, in the following quote, that a human target *"might"* be a "poacher" (or might not be?) and that however some of us might be morally repulsed by the killing of humans, the article found the outcome to be *"salutary."*

> If guards and game rangers come across anyone who might be a poacher, they now have the right to shoot first and ask questions later. Shoot-to-kill doubtless offends the sensitivity of some western conservationists, but in the Zimbabwean bush its effect was immediate—salutary (*Economist* 1997b, 96).

One could call this the dirty little secret of the wildlife conservation movement. However, it should be stated that the members do know about the actions of these organizations to interfere with governments' attempts for economic development, and they seem to support these actions. And although giving helicopter gunships to governments to enforce shoot-to-kill policies may not be common knowledge, there is enough that is blatant and so widely known that its members should know better.

Many of the efforts to preserve wildlife and conservation activities often had racist and elitist origins. This was particularly true when it involved European countries reserving areas in their colonies to be used for hunting by expatriates (Mackenzie 1988; Kjekshus 1977). These wildlife reserves were and remain reservoirs for disease vectors, such as the Tsetse fly, which threaten the life and livelihood of the poor who live on the periphery of the park.

The conservation movement in the United States early in the 20th century has received relatively uncritical praise. My school textbooks never explored the possibility that conservation in the United States "at the turn of the century was an alliance of rural and urban elites arrayed against more marginalized rural people including immigrants." Various measures prevented urban ethnic minorities from either fishing and hunting or participating in organizations which promoted them (Fox 1981, 351). Race was criteria for denying access to nature parks and public beaches (Poirier 1996, 741–742, cited in Dorsey 1998, 99 and 101). It is rarely noted that some leading conservation groups once actively practiced discrimination in their membership excluding blacks, immigrants, and Jews (Jordan and Snow 1992, 75–77; D. Taylor 1992). Some continued to do so until recent times (Dorsey 1998, 99–101).

Stephen Fox argues that "historians realize the unfairness of judging a previous generation by standards accepted now but not then." Nevertheless, he extracts some chilling quotes from American conservationist literature of the second decade of the 20th century that are painfully reminiscent of Nazi writings in later decades. Italians were "strong, prolific, persistent and of tireless energy." Their tendency was to "root out the native American and take his place and income." "Native American" did not mean American Indian. Italians were "spreading, spreading, spreading." The good guys, the "Nordics," thrived on farm work "but withered in factories and crowded cities where the dark Mediterranean races bred and prospered." The "Nordics" were doomed to lose out to the Jews with their "dwarf stature, peculiar mentality, and ruthless concentration on self-interest" (347–348). "All members of the lower classes of southern Europe" were a "dangerous menace to wildlife" (347).

The rural poor were "among the people who suffered most." As one author states it, conservation "secured liberties for some" but also "brought a significant degree of coercion for others" (Warren 1997, 49, 177, 181). Louis Warren raises some very serious questions about how conservation was carried out in the past and the implications for conservation in the present and future.

> There remain many dark and troubling questions about ... how a nation should administer public resources.... Whether a nation can act as a community, and whether in this regard a true national commons is even possible, remain to be seen (182).

Criminalizing the behavior of the local population is as old as wildlife conservation. To one author, wildlife conservation in Europe originally had "nothing to do with ecology or a love of nature. The goal of these efforts was to ensure the continuing availability of animals to be killed by sport hunters. Like it or not, it was the people who got a kick out of killing animals who laid the historical basis for wildlife conservation" (Rensberger 1977, 216). Those who hunted for necessity were scorned as taking "pot shots" or shooting to put something into the pot and satisfy his or her family's hunger. That person killed "without regard to etiquette, humanity, law or even the common decencies of life" (Schmitt 1969, 10; see also, Warren 1997, 14). In colonial Africa, there was a similar contempt for those for whom the "sole motivation" for hunting was meat (Neumann 1998, 107).

Under colonialism in Africa, preserves were created to sustain the aristocratic virtues of the "Hunt" (with a capital "H") that presumably reflected the virtues that drove English and European global dominance (MacKenzie 1987, 41; see also Carruthers 1995a, 105–108). Hunting as sportsmanship was a "ritualized act" of "upper-class identification" for the colonialists both at home and in the colony (Neumann 1998, 106). Ralph Stanley-Robinson, famed tiger hunter, reminded his hunting companions—"The object of this hunt is imperial. We are rulers here" (Cartmill 1993, 136).

The concept of the "Hunt" highlighted the conflict in values and ethical systems between Europeans and Africans under colonialism and in many ways still generates conflict in the post-colonial era. Hunting by Europeans in Africa was assumed to be benign, while the Bantu were presumed to be "*exterminating* everything in their path" (Carruthers 1994, 281). There were colonialist claims of a "wildlife slaughter" by Africans. Animals killed by Europeans were called "game," those similarly dispatched by Africans became "victims" (Neumann 1998, 107). Unless one accepts the Afrikaner myth that the Africans were late arrivals in Southern African, it is difficult to explain how Africa could have been so replete with game animals if the Bantu and other Africans were so destructive of the fauna.

Upper and middle class European values about sport-hunting and the cruelty of snares and trappings were imposed on Africans whose values were the opposite: killing for sport was wasteful and snaring was appropriate sustainable utilization of a natural resource (Carruthers 1999, 7; Carruthers 1995a, 91).

There were also cultural differences between the British sportsman who visited South Africa to hunt for trophies and the Boers, who like the Africans, hunted for subsistence and the market (Carruthers 1995b, 40–41). Those who didn't shoot for necessity were "sportsmen" and as sportsmen there were some animals you shot and some you didn't. Others were defined as "vermin" at which anyone could shoot because they were to be eliminated. Game laws in colonial Africa prohibited shooting animals from vehicles. Vermin were excepted from this rule, and in early 20th century game laws (for example in Kenya), vermin included lions, leopards and cheetahs (Rensberger 1977, 37; Adams and McShane 1996, 45; for southern Africa, see Beinart 1989a, 151; Beinart 1989b; Carruthers 1989b, 190; Carruthers 1995b, 96–97; Mutwira, 1989 253–255).

When lions were transformed into game, they had to be prodded, called "galloping the lion," to work them into a rage and to force them to charge the hunter before he shot to kill (Rensberger, 1977, 38). "Hunters often showed more squeamishness about the killing of animals than they did in the killing of Africans. Nevertheless, evolutionary ideas abound in the literature of the 'Hunt' are linked to concepts of 'sportsmanship.' Crocodiles, at a less advanced stage of evolution, were fair game for slaughter." "Waiting up for animals, concealed in a tree or a hide for example, shooting from train or river steamers, going for the leg shot to cripple rather than head or heart shot were all considered unsportsmanlike" (MacKenzie 1987, 52). These sportsmen and conservationists, or those they hired, shot or poisoned thousands of animals, "such as leopards, hyenas, baboons and bush pigs" because they were "vermin" or "pests" (Rensberger 1977, 221–222).

Many of the boards of wildlife conservation organizations today are dominated by hunters who wish to preserve the animals so that they will be there to hunt and kill and not out of any touchy-feely love of animals. Fees from permits for hunting and fishing have long supported conservation activities in the United States and could conceivably continue to play an important role in wildlife conservation around the world. But it is important that this not be a hidden agenda both for the political discourse and for those who support these organizations with their voluntary contributions.

Conservation and CITES

In 1900, the colonial powers created and signed the Convention for the Preservation of Animals, Birds and Fish in Africa. "It was the first international conservation treaty and ... it became the basis for most of the wildlife legislation in Africa, and the forerunner of the Convention on International Trade in Endangered Species (CITES), which came into being seventy-three years later and is the most comprehensive treaty today" (Bonner 1993a, 39). A key phrase in the 1900 treaty was the expressed desire for the colonial powers to save, in their possessions, "various forms of animal life existing in a wild state which are useful to man or harmless" (40–41). Thus, as previously noted, they were free to kill and even exterminate species deemed to be vermin. In fact the driving force of the first treaty was not conservation in the current use of the term but the creation of game reserves that "could be used by hunters for sport and trophies" (Steinhart 1994, 60). The colonial governments carried out their policy by creating wildlife preserves,

much as the royalty had done earlier and for many of the same purposes. One of the purposes was to control the sources of ivory, as the
export of ivory was a major source of revenue for the colonial governments. This is in sharp contrast to the same former colonial powers
who now impose their will on the independent African governments
not to export ivory.

There is an interesting but not unexpected double standard operating here. When developing countries attempt to define the resources on
the ocean floor or the accumulation of knowledge as a "heritage of
mankind," a loud howl is heard in developed countries about interfer-
ence with individual freedom and private property. But these same
countries are quite willing to declare the flora and fauna of Africa or
other developing countries to be a global "heritage" that must be preserved through international action with or without the consent of the
local population or their chosen representatives (Adams and McShane
1996, 233; LaFranchi 1997). An even more tragic double standard is
the fact that so many developed countries that promote African flora
and fauna as a global heritage do not accord the same right to the
cultural heritage of Africa. A quarter century after the UNESCO
Convention on the means of prohibiting and preventing the illicit
import, export, and transfer of ownership of cultural property, "only
two countries in the industrialized Northern Hemisphere—the United
States and Canada—have confirmed their ratification with a specific
set of laws," while most of the other major industrial countries—
Germany, France, Japan, and United Kingdom—have failed to even
ratify it (Brent 1996, 75). Such failure means that money can be made
in trafficking in stolen cultural treasures without fear of criminal and
other legal sanctions while trafficking in wildlife can (and should
result) in criminal penalty. It therefore not only guarantees the continued massive looting and destruction of the cultural heritage of Africa,
it also guarantees the destruction of archeological sites and what we as
humans can learn from them. This is truly a loss for all of humanity.

Developed countries impose wildlife conservation on less developed countries to placate the "green" movements in their countries
without themselves having to make any sacrifices. We are told that,
regarding the earth and the environment "we are all in this boat
together." True, but this ignores the fact that some are in the cabins and
some are in steerage, and those in the cabins are placing the burden of
keeping the boat afloat on those in steerage.

In the late 19th and early 20th century imperialists often justified
their colonial conquest and subjugation in terms of a "dual mandate."

The presumed "unexploited" resources of Africa were to be developed for the benefit of the local population and for the benefit of all of mankind. In effect, the claims of colonialists and modern conservationists are predicated on virtually identical philosophical premises. Brazil was long faulted for not developing the Amazon region and felt that others coveted this region for their own exploitation. Now Brazil finds that it is being faulted by the same people for seeking to develop the Amazon region. Again, Brazil feels that others are seeking to control its destiny.

An even more outrageous double standard is the fact that the affluent can get a license to shoot an otherwise protected species in Africa and can import the results of their kill, such as a "trophy" ivory tusk, into the United States. Or one can go to an auction in Southern Africa and buy an elephant (Bakker-Cole 1995, 6). If the same elephant were shot by an African government as part of a culling process for conservation and the tusks were made available to local carvers, their product could not be imported into the United States under the special exemptions for trophy tusks to the African Elephant Conservation Act (Bonner 1993a, 270; Ross 1992, 389).

Africans were and are prohibited from hunting in these preserves and therefore were denied what had been an important source of protein in their diet. In some instances, they were effectively prohibited from being in the preserve at all (Beinart 1989, 150; Carruthers 1989b, 189; Carruthers 1995a, 99; Carruthers 1995b, 26, 64, 138–139). When famine struck, they no longer had a hedge against it (MacKenzie 1987, 42–43, 50, 57). During South Africa's drought of 1913, "many resident Africans were dying of starvation" but were still not permitted to hunt for survival (Carruthers 1989b, 198; Carruthers 1995, 93). Not being sportsmen by definition, Africans were defined as "evil, cruel, poachers" who wanted to live off the land rather than engage in honorable wage labor (Carruthers 1994, 271–272). In the 1896–1897 drought, destitute Boers were prevented from using the game in protected areas of South Africa (Carruthers 1995b, 80–81).

Virtually everywhere in the Third World today, the local population is seen as the biggest "threat" to wildlife preserves (Ghimire 1994, 198). This is not only true for the attitude toward "poachers" but toward cultivators as well. A regional director of one of the major wildlife conservation groups considered the "transformation of formerly natural ecosystems for human purposes" as the most severe environmental problem facing Tanzania (Lamprey 1992, 10). We still speak of the "senseless slaughter" of African wildlife. "The hunting of food or of animal products that can be exchanged for money by impoverished Africans is hardly 'senseless.' It is, in fact, quite sensible, as it is the only source of livelihood" (Rensberger 1977, 219). Hunting by Africans has also historically been for the raw materials to make the artifacts that were an essential part of their cultural practices (Bonner 1993a, 43). There were, and remain today, many other conflicts of interest between the local African population and the imperatives of the game park. In many cases they are a reservoir for diseases which threaten local livestock.

Animal Rights and the Rights of Humans

In the name of conservation, there are those who in essence bribe Third World governments to interfere with and criminalize the behavior of local inhabitants (Pearce 1990a, b, c). Many promote international boycotts of sealskin products, thus depriving Inuit hunters of their traditional livelihood, by using publicity films that involved "ruthless trickery" (Matthiessen 1995, 74). The antifur crusaders are obviously not as concerned that the bans on fur imports are adversely effecting the traditional cultures of Cree, Ojibwa, and Sarnia Indians whose livelihood is from hunting mink, fox, and beaver (Payne 1997).

Many authors and journalists do not see the contradiction between writing on the traditional hunting behavior of a group and describing them as "poachers" when they hunt on what was once and in their minds is still, their traditional lands. In an article on Kahuzi-Biega National Park in the Congo ("where Dian Fossey first saw wild gorillas"), the park is referred to as "once the home of pygmies who hunted and collected wood and plants for medicine." The next paragraph refers to a frequently arrested pygmy hunter and gatherer—62 year old Bulabi Lubaga—as a "poacher" even though he was only caught "collecting honey and trapping monkeys to survive" because, as he states it, he "was hungry." The author was sympathetic to Bulabi Lubaga and did admit that he was "only a bit player in a poaching problem;" nevertheless, by using the term poacher, there was an implied criminality in traditional behavior in an indigenous person's homeland (Fisher 1999; see also Jenkins 1999).

Conservation and Hunting

There are some endeavors so overwhelmingly right that we sometimes fail to examine who is promoting them, what their motives may be, and how they are carried out or who bears the cost. Conservation of habitat is an obvious necessity such that no intelligent, informed person in the 21st century could oppose it on principle. So is preservation of biological diversity, cultural diversity, and all our different heritages. Needless to say, if failure in these and other areas of the human activity threaten the long-term enterprise of being human and the civilizational endeavor, then policies should be framed and actions taken to sustain culture and humanity. Possibly these are the policies about which we should be most critical, to make certain that the goodness of apparent intent is not a mask for actions that result in undue benefit to some and harm to others. Doing what is right in the right way is imperative. Otherwise, criticism of the way conservation (or a similar activity) is carried out might become a basis for criticizing the inherent necessity for the action itself.

Wildlife Conservation and African Independence

Colonial governments enforced the hunters' exclusion of everyone else from the game parks. Now the independent governments work with international conservation agencies that protect the parks for the animals, for the tourists, and for the foreign exchanges that they bring. It was the colonialists who were either directly or indirectly responsible for the species that had become extinct in Africa in recent centuries; still, Europeans assumed that Africans would destroy the environment unless they were prevented from doing so. In the late 1950s and early 1960s, as African countries were becoming independent, many in the West were fearful that these newly independent governments could not discharge their responsibility to protect wildlife. This fear was the impetus for the formation of conservation organizations in the United States and Europe to protect African wildlife (Bonner 1993a, 57, 64, 176).

Basically, Americans and European do not trust Africans to manage their own environment. Even when African countries have sensible private sector programs of game ranching that offer genuine opportunities for preserving threatened species and profitably involving the local population, many conservationists oppose these programs, such as "Campfire." Campfire is an acronym for Communal Areas Management for Indigenous Resources, which was begun in Zimbabwe in 1989 "as a way of reducing complaints about marauding animals in villages next to national parks. It allowed indigenous residents to benefit from game roaming on their traditional lands" (Wells 1997; see also Peterson 1994; Taylor 1994; Kasere 1996; Daley 1997a, 1997b; Nyoni 1997; Postrel 2000; Adams and McShane 1996, 178–183, 257–260, also 153 for a similar program of local involvement in Botswana; for Ghana, see Ameyibor 1997).

Many conservation spokespersons from developed countries promote the value of local involvement in wildlife conservation. Yet they oppose measures such as the change in CITES allowing the culling of elephants and the export of ivory from southern African countries, thus depriving local villagers of the major potential source of income from the game park. One Third World spokesperson, Dr. Mostafa Tolba, said at the 1992 CITES meeting:

There are loud cries from a number of developing countries that the rich are more interested in making the Third World into a natural history museum than they are in filling the bellies of its people (ART 1997, 10).

Dr. Tolba then went on to question whether those supporting CITES see its "principal role as preserving species or utilizing them for sustainable development?" (10).

A CNN correspondent succinctly stated the benefit to wildlife conservation of making it profitable to the local inhabitants: "If it pays, it stays" (Hanna 1997; P. Brown 2000; on other cooperative efforts for sharing the benefits of game preserves in Southern Africa, see Hammond 1997; for opposition to a policy of making wildlife pay for itself, see Macleod 1997a, 1997b). One effective way to preserve species and habitat and secure local cooperation is to be "politically incorrect" and promote sport hunting, with the local inhabitants getting their share of the revenue. According to research in Southern Africa, hunting in Campfire-type programs "creates jobs and training opportunities for local residents" and provides more revenue than other forms of tourism (Koro, Ovejero and Sturgeon 1999, 53).

The Campfire program is summed up as follows:

Basically, it assesses game populations and awards permits—of an elephant or two, for example—to local communities. Villagers can harvest the animals themselves and can, say, use them for food. Or, as in many cases involving elephant permits, they can sell the permits to safari operators who guide big-game hunters into the area (Wells 1997, A11).

The famed conservationist, Richard Leakey, makes some sensible comments about the dangers of conservation based on "private reserves run largely by Caucasians" which he likens to "sitting on a time bomb waiting for it to bang." He goes on, saying that "we mustn't make the mistake of excluding people from their land. One way to soften the inside/outside divide is to get into community involvement. This has become fashionable now" (Macleod, 1997a). Leakey then adds an important caveat that appears to undercut the rest of his argument.

> But having been a champion of sharing revenue with communities, I am now opposed to it. Poor people cannot be expected to make the right judgments about the protection of species (Macleod 1997a).

In other words, Leakey doesn't want the local community excluded from the conservation process but neither does he want them "involved in managing national parks." And Leakey is definitely opposed to any scheme that makes a species pay—through eco-tourism or hunting licenses for example—for its survival. "It is *Homo sapiens* who must pay. The point is that species must stay so we must pay" (Macleod 1997a). All of which is fine, but on Leakey's principle, the *Homo sapiens* who will end up paying will be those least able to afford it. In this case, it will be Africans.

Leakey's point about not excluding people from their land needs to be emphasized. The Western conception of conservation as applied in Africa was to exclude people from the land (Carruthers 1999, 2). Europeans in Southern Africa, colonial administrators and settlers alike often predicated their game conservation policies on "purists parks" ideologies of a pristine "old Africa" (Ranger 1989, 230, 232, 230). "Nature conservation is thought to be intrinsically good" and parks "are generally considered to be morally sound" (Carruthers 1995a, 1). The legitimate question is "whose nature?" or "whose heritage?" is being preserved, African or that of the *external* conservationists (Beinart 1989, 156; Ranger 1989).

What is "nature" to the expatriate conservationist is in fact "heritage" to those who live there. "In the Imperial European conceptual map of the world, Europe was culture and the colonies were nature" (Neumann 1998, 32). Preserving nature and protecting cultural heritage are not always compatible goals. "Rather than being 'unspoiled benchmarks,' most of the continent's protected areas have been created out of lands with long histories of occupancy and use" (4). The real issue is not one of saving the earth, but whose vision of Africa should govern policy actions, that of the Africans or the expatriates? Neumann quotes a "British expatriate ecologist serving as an official of an international conservation organization" who, on seeing a park with a landscape devoid of humans ("There were no obvious signs of human life or activity in the landscape, save the dusty ruts."), proclaims that "this is the way Africa *should* look" (1). This, in Neumann's mind raises a number of critical questions:

> Who decides what Africa "should" look like? Where and how have ideas of the pleasing African landscape been constructed? What does this landscape vision mean for African peasants and pastoralists living and laboring there? To what degree and in what ways do they resist this vision? (1)

Fortunately, Carruthers sees new conservation policies emerging in South Africa and other independent African countries where "protected areas are now being used as tools for rural development and capacity building rather than being merely tourist playgrounds" (Carruthers 1999, 15). This trend exists and it should be supported both by conservationists and their critics.

Wildlife Conservation and the Ivory Trade

When governments in Southern Africa achieved success in protecting elephants to the point that there were too many, thereby endangering the habitat, conservationists still opposed their selling the ivory that resulted from the ecologically necessary culling (Bonner 1993a; Concar and Cole 1992; Makova 1997a and b; PANA 1997). Failure to cull wildlife, particularly elephants in protected areas, will lead in time to their expansion beyond the carrying capacity of the land and therefore cause severe environmental destruction. In times of drought or other climatic fluctuation, the already stressed environment will be further degraded, resulting in major losses in animals. Because of a drought in the late 1960s in Kenya's Tsavo National Park, "at least 9,000 elephants and several hundred rhino died of starvation, the elephants having destroyed not only their own food supply but the rhino's also" (Bonner 1994a, 60).

Many critics argue that the international campaign to "save" the elephant was not because they were in danger of extinction but because it was an effective way of raising money for wildlife groups (Showers 1994, 42; Bonner 1993b, 18). This often meant that actual data on the wildlife situation in Africa was distorted to make the situation worse than it was (Kreuter and Simmons 1994, 43). Among the distortions was the inflation of the figures for the decline in the elephant population by including 300,000 elephants in Zaire that never existed (Pye-Smith 1999, 16). "Elephants' doom was money in these groups' pockets. Successes do not seem to have the same result" (Adams and McShane 1996, 76). And "few of the dollars thus raised have reached Africans actually working in elephant conservation, because the issue was stopping the ivory trade, not hands-on-programs on the continent" (60).

Using the CITES treaty, the developed countries succeeded in imposing a ban on the ivory trade (except for the previously noted "trophy" kills), even though the ban was opposed by a majority of African member countries, including those with the most successful conservation programs (Chadwick, 1992, 466–467). It is interesting to note that most of those voting for the ban on the ivory trade had never been to Africa (Adams and McShane, 1997, 65).

> When the Convention on International Trade in Endangered Species of Flora and Fauna (CITES) ban was being formulated, there was no consultation with African governments or local wildlife protection institutions. The effects of the ongoing regional wars on wildlife populations or local governments' capacity to enforce wildlife protection were also not discussed. Rather than working with the national organizations to strengthen their resources and programs, the northern hemisphere chose to implement unilateral policies which profoundly interfered with national schemes (Showers 1994, 43).

At the June 1997 meeting of CITES, the Southern African countries sought an exemption allowing them to cull elephants and to export ivory under very carefully controlled conditions (Mapininga 1997; Wilson 1999; Mehra 1999). The continuation of the ban on ivory exports was supported by a number of African countries (Mulenga 1997). Many environmentalists complained that the meeting was being held in Africa. Presumably being close to the situation and being involved in the outcome introduces an unacceptable bias. Only those distant from the situation who bear none of the burdens of enforcement can be "responsible," "objective," and "global" in the decision-making process.

True to earlier form, on the first vote at the 1997 CITES meeting, the Southern African states failed to get the two-thirds majority necessary to overturn the ban. A majority of African states voted to lift the ban, while opposition came mainly from developed countries. On the second and final vote, a two-thirds vote was achieved for a compromise that allowed the three Southern African countries to export ivory solely to Japan in a tightly regulated trade. At the April 2000 meeting of CITES, there was a debate over whether Southern African countries should be allowed to continue to cull and sell ivory or whether the previous permission was to be a one-off sale. The compromise was a two-year moratorium on ivory sales while mechanisms were put in place for monitoring and controlling the process, at which time the resumption of sales would be considered a possibility (Jenkins 2000; Mwangi 2000).

Wildlife and Human Life

"Most conservation agencies are paramilitary armed and uninformed organizations, in which the majority of expenditures is devoted to law enforcement and public relations" (Bell 1987, 88; see also, Neumann 1998, 5–6). In many instances, game wardens are ex-servicemen who a few years earlier were out hunting Africans fighting for independence.

> From the first, there has been an association between game parks and military men all over Africa. In part this is for the obvious reason that ex-soldiers often make good game wardens, accustomed to the outdoor life and trained in the use of weapons (Ellis 1994, 55).

The policy in most parks where humans, other than tourists, are spotted is to shoot on sight and often. "Shoot to kill" is explicitly stated. "Helicopter and tommy gun combat teams patrol through an African valley in search of 'poachers' using equipment that is often provided by international conservation societies and donor development organizations" (Knox 1990, 47–48, 50; see also Vollers 1987; Bonner 1993a, 78). These organizations have provided African governments "more rifles, bullets, helicopters, vehicles and equipment to conduct war." As Bonner adds, despite these measures, "poaching escalated," which suggests that the methods may be ineffective as well as immoral (Bonner 1993a, 19, see also 18, 78). These same groups apparently also "provided funds to armed antipoaching units in Namibia, set up by a clandestine and proscribed operation, run by a team of British mercenaries ... to infiltrate the illegal trade in rhinoceros horn" (Bogan and Williams 1991).

The result has been a slaughter in Africa which few in the developed countries notice or about which anyone even seems to care. The previously noted helicopter "helped to trap and kill 57 poachers within three years" (S. Armstrong 1991, 55). In Kenya, "more than 100 poachers were shot—legally—last year" (1990) (Barden 1991). Being "legally" shot by these Rambos of the Rift presumably was supposed to be a consolation to those who were carrying out traditional hunting activities. "In the past five years, more than 60 'poachers' and at least one scout of the Zimbabwe National Parks Department have been shot dead in fire fights" (Knox 1990, 48). One organization promoting "cultural tourism" informed its members that "Zimbabwe lost 27 rhinoceros to poachers in 1990, but its new conservation program, initiated last year, allowed rangers to shoot to death 28 poachers" (*New Yorker* 1991b, 116; ICOMOS 1991, 3). For many who remember the Vietnam War, statistics on the "body count" of "poachers" conjures unpleasant associations.

In another instance of killing intruders, without first obtaining complete information, a campsite was mined. The heat from the fire of the returning "poachers" set off the mine. "'We came back and found a biltong tree,' chuckles Edwards" (Steve Edwards was the game warden). An American environmentalist adds, "Biltong is beef jerky. Is this modern environmentalism?" (Knox 1987, 48).

> As a newcomer, it was difficult putting Edwards' fighting words together with my Edenic surroundings. An orange sun was simmering into the river upstream from us. Two elephants splashed ashore on the bank nearby. Fireflies blinked. And we sat talking about triangulated gunfire, claymore mines and the biltong tree. The wilderness has become an armed fortress (48).

Unfortunately other environmentalist and environmental/conservation groups are not as sensitive to the rights of local populations as Knox is. Many of these groups encourage and applaud actions which lead to the death of the local inhabitants (Bonner 1993a, 66–67). Many of those killed or captured are local peoples who are carrying out what were to them traditional practices. As one local hunter stated, "how can you tell me I don't belong in a place where I have lived my whole life?" (Adams and McShane 1996, 122, 127–130). Adams and McShane tell of an African park employee, shot and killed by a white security guard, who was not punished for it (228–229). But when a white park manager or tourist is killed, it becomes internationally known.

It is more than a little ironic that some of conservation groups working with governments in Asia or Africa have military sweeps through villages to confiscate any weapons used for hunting, "arms surrendering" ceremonies, and require a pledge not to hunt as a condition for habitation on the periphery of a park (Ghimire 1994, 208). This particular program is, incidentally, often touted by conservation groups as being one that has successfully involved the local population in environmental protection. The hunters in these conservation groups would howl to the heavens if their hunting weapons were confiscated.

Krishna Ghimire gives more examples of populations displaced for park creation. These include "thousands" of tribal people evicted in India in the 1970s in order to create tiger parks, and the "expulsions of the Rendile from Sibiloi National Park in Kenya, the Ik from Kidepo National Park in Uganda, and the Masai from Serengeti National Park in Tanzania ... (and) ... 1,100 villagers living in Korup Park in Cameroon." "People are usually transferred to entirely different socioeconomic or climatic zones, or given very small land plots, forcing many of them to re-enter forests for 'unauthorized' cultivation and extraction of forest products" (Ghimire 1994, 223; see also Kamuaro 1996).

In Park Montagne d'Ambre in Madagascar, "local people are prohibited from entering the park for any purpose and are liable to be arrested and fined if found inside park boundaries.... Many of these areas were previously used by local people for growing fruit, vegetables, or *kat*, or for grazing livestock." This was also the case for Ranomafana National Park in Madagascar (Harper 2002). The closure of the park to all local uses has meant that local people are not even able to collect such items as dried wood, nuts, berry shoots, and medicinal plants, which are renewable and whose removal would cause no serious forest degradation (Ghimire 1994, 220; see also Neumann 1998, 5). In some instances, such as Manas Wildlife Sanctuary in India, resources such as grass, which the local people could have used for thatch or animal feed, is burnt by Sanctuary authorities (Ghimire 1994, 224). Skukuza, the name of the rest camp in Kruger National Park in South Africa, means "he who sweeps clean" in Tsonga, a name given by "tribesmen forced to vacate their villages so that the reserve could be built" (Koch, Cooper, and Coetzee 1990, 17; see also Armstrong 1991, 54; Carruthers 1999, 5). In reference to Pilanesberg Reserve, we learn that:

Inside its game fences are the remains of homesteads whose residents
were removed in the early 1980s to make way for the park. People living off the land adjoining the reserves are denied access to the trees,
roots, grasses and herbs that as food or medicine are an integral part of
their lifestyles. Hunting and fishing—essential means of obtaining
food—are harshly punished (Koch, Cooper and Coetzee 1990, 17).

Nature parks are still being created in Africa and around the world,
with the indigenous population either being expelled or forced to live
in terms of someone else's definition of their cultural lifeways. In late
1997, there was an announcement for the creation of an $800 million,
580,000 acre game park in Mozambique. No mention was made of how
many people would be displaced. The claim was made that people
would "not be forced to move" but instead would be "persuaded to
move to villages outside the park" where they could "benefit from jobs,
development and money from the project" (Reber 1997; see also
Sayagues 1998, 1999a, b, c; Koch 1996, 1997). In other discourse in
Mozambique on conservation issues, the civil war's "displacement and
depopulation" of previously inhabited areas was seen as a "window of
opportunity" for the creation of an "elephant-centered park." Zerner
finds that although the "project brief does acknowledge" many rights of
the local community, it obscures critical issues, weighing the rights of
large mammals against "the rights of local communities to freely return
to their lands and to practice forms of livelihood of their own choosing"
(Zerner 1996, 93).

Post–World War II ideas on wildlife conservation in Africa had origins other than the colonial countries. Bernhard Grzimek was zoology
curator at the Frankfurt zoo under Hitler. He is famous for the book and
Oscar-winning film, *Serengeti Shall Not Die* and honored as the father
of modern African conservation. The British author George Monbiot
quotes him, arguing that "a National Park must remain a primordial
wilderness to be effective. No men, not even native ones, should live
inside its borders" (Monbiot 1999; see also Grzimek, 1961). This is
Nazi purism in its purist form. Grzimek ideas were in line with already
established British colonial wildlife conservation policies in the
Serengeti. One conservationist is quoted by Neumann as arguing that
"the interests of fauna and flora must come first ... those of man and
belongings being of secondary importance. Humans and a National
park can not exist together" (Neumann 1998, 136; for a critique of the
imposition of this cultural ideal on others, see Croll and Parkin 1992).
"The idea that the environment is something separate from the people

that inhabit it is a distinctly Western concept, and is not a view shared by all societies" (Harper 2002).

Ironically, the people who presumably have had a lifestyle that does not destroy the environment are most likely to have their land taken from them. Yet, the people of the area are not "seen as part of the total ecology of the area ... but as problems, obstacles who could be ... frequently ... moved elsewhere so that the environment could be 'preserved'" (Koch, Cooper, and Coetzee 1990, 21). Beginning in the days of colonialism, Africans have been seen as enemies of conservation, intruders and ravagers of the habitat, all of which therefore warranted European "custodianship" (Carruthers 1995a, 90; Carruthers 1999, 3). African "poachers" were portrayed as "the most bloodthirsty, cruelest and most ruthless of the earth's inhabitants" which was consistent with the long-held European view of Africans as cannibals and bloodthirsty barbarians (Carruthers 1994, 281; Carruthers 1999, 9).

A more recent variation of this same theme can be found in the assertion that those "who depend upon their surroundings for their living are not in a position to take care of their environment" (Brockington and Homewood 1996, 101). It is precisely this belief that has given affluent Westerners a license to work with colonial governments (and feel morally superior doing so), and then later with the independent governments that followed, to force people from their ancestral lands. Lands which they had obviously long shared with the wildlife, since the animals were still there. In one way or another, under colonialism or post-colonialism, the local population has in some way to be dehumanized in order to justify depriving them of their land and livelihood.

> It was necessary to dehumanize the Africans who lived and worked in the virgin landscape so that reality would fit within the vision. "Primitive" Africans were often simply regarded as fauna ... The possibility of protecting them could therefore be given serious consideration (Neumann 1998, 128).

Under colonialism (and to some extent since then), parks were not seen as a "symbol of national pride" but were instead "perceived as part of a governmental structure from which they had been systematically excluded." For the colonial government, the parks came to be seen as another means of control over the African population in the area (Carruthers 1995b, 176). Africans were (and often still are) living on the "other side of the fence" in "overcrowded, degraded, and unattractive rural and urban environments" (Carruthers 1995a, 89). For the white regime in South Africa, Kruger National Park was "represented as a remnant of the 'wilds' that Afrikaners had struggled to tame" (Neumann 198, 32).

In recent years, environmental groups have "discovered" indigenous peoples who would be displaced by the construction of a dam and irrigation project, which would provide electricity and water to grow crops. Yet these groups not only have remained silent on the displacement of populations for environmental projects, but many have been active participants in the displacements. Calling these actions hypocrisy would be a gross understatement.

An environmental scientist, who is an advocate for moving indigenous populations, criticizes other environmentalists who *don't want to displace* indigenous populations as being unrealistic, harboring romantic ideas about "noble savages" and "dooming these people to sustainable poverty." This scientist is with a conservation organization that has 250 projects in 52 countries. He thinks that it is "a 'desirable goal' to move up to 6000 people from the Nagarhole National Park in Southern India in order to preserve 40 Asia tigers," though he "stresses" that they should be "encouraged, not forced to leave" (Edwards 1997, 15). Another scientist from a society that "oversees 160 projects in 44 countries" argued that "relocating tribal or traditional people who live in these protected areas is the single most important step towards conservation." To him, tribals "compulsively hunt for food" and are thereby competing with tigers for prey (R. Guha 1997, 17).

Dian Fossey: Saint or Sinner?

Dian Fossey seems to be showing a preference for animals over people in the diary entry that opens a eulogizing biography of her:

> From my childhood I believed that was what going to Africa would be, but by 1963, when I was first able to make a trip there, it was not that way anymore. There were only a few places other than the deserts and the swamps that hadn't been overrun by people (Mowat 1987, 1).

It seems Dian Fossey is saying Africa would be a great place if there were not so many Africans. Even worse, Fossey was known to call Africans "wogs" or "apes" leading some to believe Fossey felt that, of the two large primates in Central Africa —gorillas and Africans— gorillas were to be protected at the expense of Africans. However, long after Fossey's death some still believe that, in her words, "the last great communities of gorillas living in the wild" are Africa "at its best" and that we can thank her for their still being there.

A TV anchor for the American network with the largest audience has very little time for Africa and its people on the nightly broadcast unless they act brutally against whites, becoming "Africa at its worst" (Lobe 1999; see also, Ramaphosa 1999). Killing tourists is news, as it should be, but killing Africans as part of wildlife conservation is not news. As Susan Moeller aptly puts it: "It's not called the 'Dark Continent' for nothing" (Moeller 1999).

Those of us who found fault with the book about Fossey are even more critical of the film "Gorillas in the Mist" about Dian Fossey's work in Africa. An argument can be made that both the book and the film are racist. In the film, an American, who admittedly has no expertise in African cultures or primate behavior, goes to Africa to save the mountain gorilla. In fact, Fossey was chosen because Louis Leakey believed "that the best person to study primates in the wild was a scientifically untrained woman" (Adams and McShane 1996, 1860). What does this intruder in the mist proceed to do? She routinely destroys the traps that the local population has historically used to catch small animals that provide protein for a nutritionally deficient diet. Later in the film Fossey requires those who work with her to do the same. She desecrates a burial ground. She kidnaps and terrorizes a young boy and in the process treats the local beliefs as superstitions. Fossey strips a man of his amulets, conducts a mock hanging and, as she correctly stated it, strips his manhood also. Much that was not shown in the film is equally abominable. According to Adams and McShane, Fossey, "a tall, hot tempered woman," intimidated the local population.

> She would torture those poachers she could catch, whipping them with stinging nettles, putting nooses around their necks, kidnapping their children and burning their possessions. She also waged a psychological war, which included terrifying suspected poachers into believing that she was a sorceress capable of casting spells (194).

In the movie, she burns the villagers' huts, while she roughs it in a dwelling with electricity (whose source is not indicated) to run her hair dryer, tape player, and electric lights. Everything is provided her at a standard of living that is beyond the wildest imagination of local inhabitants whose livelihood she threatens (Nash and Sutherland 1991, 115–118).

What is most outrageous about the film is that so many viewers have accepted its central thesis.

Fossey believed that only she could save the gorilla. Instead, she became the biggest threat to the gorillas' survival (Adams and McShane 1996, 195–196). From the time of her death to the outbreak of violence in Rwanda, the gorilla population increased as a result of a local initiative, the Mountain Gorilla Project (185–186, 199–201, 205). Tourism, which helped to pay for the protection of the gorillas, was opposed by Fossey who called them "idle rubberneckers," and even fired shots over the heads of a group of tourists (198).

While this judgment of Dian Fossey may seem overly harsh to some, she would probably not want it any other way. She was interested in results, and not in the favor or judgment of her detractors. She was a person of great passion who was willing to go to any extreme to save the gorillas. Knowing the danger of her situation, she paid the ultimate price for her passion. As with many who have followed her in a passion to save a diversity of life forms, we can respect their intent, but fear their extreme passion which too often ends up being destructive, and in Fossey's case, self-destructive.

Placing the burden for conservation on the poor in countries that have not yet depleted their flora and fauna the way we in developed countries have, is as prevalent in our own hemisphere as well as in Africa and Asia. An article in *Audubon Magazine* titled "Peace Is Hell!" stated that "This may sound crass, but peace is going to be a disaster for Nicaragua's environment." The fear was that farmers and loggers would return to reclaim or colonize lands that were unsafe during the recent war there. The solution in yet another voice in the article was to receive "international assistance to patrol and stabilize" the area. The famed trade-off between guns and butter here means that we should give guns to the overseers for authority so that they can use them against campesinos who "are filtering back to claim the fincas they abandoned when the trouble started." Ironically, we are told that the environmental damage from low intensity warfare was "modest." A not entirely facetious interpretation of the article might be that promoting low intensity warfare in environmentally critical areas might be a very effective way of furthering conservation objectives, although few people would be willing to draw that conclusion. World Wildlife Fund (now called World Wide Fund for Nature) runs advertisements to "fight poachers" and "hire guards" (Wille 1991). They have since changed the ads to "fight poachers, hire anthropologists."

The local people who are often displaced in order to create a park are sometimes allowed to remain as part of the "natural" or "pristine" habitat, provided that they survive by only using an obsolete traditional technology and "provided that they live 'traditionally' as defined not by them but by the local white nature conservator" (Gordon 1992, 183; Perlez 1991). This means that the Maasai can herd cattle but not grow maize, while the San (Bushman) cannot even raise livestock, though these proscribed activities are part of their traditional behavior (Bonner 1993a, 179, 183–184; Monbiot 1994, 93). To the true purist, even the Maasai with "their herds of economically worthless cattle are a threat to the environment" (R. Guha 1997, 14). By cultivating in Ngorongora Crater, the Maasai had earlier failed to "live up to European stereotypes" of what their traditional behavior should be. This was explained as being the "result of the Maasai having become 'much adulterated with extra-tribal blood'" (Neumann 1998, 136–137).

Africans whose behavior did not fit with British preconceptions of "primitive man" could not be allowed to remain in the national parks, regardless of their claims to customary land rights (128).

Neumann speaks of the strange "interpretations of African culture" and the "legal logic" which allowed the British colonialists to expel Maasai (and similar groups elsewhere) from their traditional lands such as Ngorongora Crater.

[s]ince we know that the Maasai do not cultivate, any cultivators in the crater must be non-Maasai, and since no non-Maasai may live in Maasailand without a permit from the Native Authority, they are there without legal rights (137).

Neumann refers to the "mythical vision of Africa as an unspoiled wilderness, where nature existed undisturbed by destructive human intervention." This "European conception of unspoiled nature" was part and parcel of the European "concept of primitive human society," all of which "had more to do with European myths and desires than with reality" (128). In other words, "Eden had been rediscovered" (32).

In other parts of Africa, the "Garden of Eden" of abundant wildlife and sparse population was often the result of crashes of human and livestock populations as a result of colonial conquest and diseases that had recently been introduced (Bell 1987, 8–9).

In some instances, environmentalists who are unable to prevent construction of a dam, have successfully demanded that a wildlife sanctuary be created to compensate for the submerged forests resulting from the impounding of water by the dam. Such is the case with the Shoolpaneshwar sanctuary in Gujarat, India, as compensation for the areas flooded by dam construction on the Narmada River. The 40,000 tribals who were allowed to stay in the sanctuary now "feel imprisoned there, and feel that their rights come secondary to those of plants and animals." Of course, they were expected to be traditional as others thought necessary in order to preserve the integrity of the sanctuary. "The tribal people in Shoolpaneshwar were forbidden from improving roads, which meant their access to health and education facilities were severely restricted. Local laws also forbade the installation of hand pumps, electricity, or any activity that could impact negatively upon the environment" (Craig 2001).

Conservation in Belize: A Case Study

It is clear that there is a plethora of material—articles, books, newspaper accounts, etc.—on the frequent, near total, disregard by environmental and other groups of the rights of indigenous populations around the world. None of the studies cited previously explore this topic in such depth and detail for a particular area as does Anne Sutherland's book on Belize (Sutherland 1998). In many respects, Sutherland's work and the previous citations complement and reinforce one another. The studies around the world demonstrate that Sutherland's Belize is not an isolated case, while the Belize inquiry more clearly delineates the driving ideology that rationalizes depriving local populations of their rights and the complex realities and motivating forces of the international structures and organizations involved in preserving habitat and wild flora and fauna at the expense of people. Sutherland explores the history of the areas set aside for conservation, the forces operating when they came into being, as well as the current conditions under which they are operating.

"Authenticity" is advertised as part of the experience of visiting Belize, though what it means to be authentic is never really explained (Sutherland 1998, 107). Authenticity is a feeling, not a fact. "This recent 'discovery' of Belize by foreigners as a tourist mecca and destination for the 'authentic' experience could only happen because in the past Belize was underdeveloped, underpopulated, and unknown" (93, see also 123). The presumed authenticity of Belize is what attracted environmentalists in the 1980s and has made it a destination for eco-tourism in the 1990s (99–101).

Sutherland describes the ecotourists and conservationists as "the new missionaries" (119–142; see also Lobe 1999). In Guha's term, they are "green missionaries ... who want to protect the tiger or whale for posterity, yet expect others to sacrifice" (R. Guha 1997, 19).

> In Belize today, the new globalized Belize of the twenty-first century, the missionaries of the past, Christian evangelists from North American sects, have been replaced by the new missionaries, the environmentalists. Armed not with bibles but with ecological fervor, they have obtained 40 percent of the land mass in Belize and reserved it for animals, fish and archaeological sites (Sutherland 1998, 120).

Sutherland gives a particularly egregious example of environmentalist insensitivity in the creation of Cockscomb Jaguar Preserve, the "world's first jaguar preserve." The Maya had long lived in the area, coexisting with the jaguar, and had befriended the particular conservationist who came to study the jaguar. All the jaguars that the conservationist (with the help of the local population) had captured for study died. He concluded that "the Maya must be moved off the land in order to 'save' the jaguar and set up the Cockscomb preserve" (Sutherland 1998, 119; see also Rabinowitz 1986; Line 1999). As Sutherland observes, environmental groups "advertise the Cockscomb Preserve as an example of eco-tourism successfully replacing formerly 'destructive ways of making a living'." They tout the fact that the sanctuary director was formerly a teacher in the local school. (Sutherland acerbically comments: "The sacrifice that the Mayan community made for the park, giving up their homes, their land and their traditional subsistence farming—in return for one job as sanctuary director—seems not to be part of the transnational environmental consciousness" (120).

After surveying the many habitat reserves on land, on the coast, and on the cays and reefs, Sutherland clearly states that "virtually everyone," Sutherland included, "agrees that there is a need to preserve the complex and delicate reef system" and of course to save other endangered habitats (129). The issue for her, for this author, and for others, is not whether conservation is necessary, not whether species diversity should be protected, but how this is done and whether the rights of the local population are respected and protected.

All of this preservation activity, while admirable, has raised a pressing question in Belize today. What about the people living in the now preserved areas, where their traditional subsistence activities have become criminal acts? (130).

Sutherland asks the question "for whom are all the reserves in Belize? Are they being established for Belizeans, for the good of the natural environment (a higher power than any human good), for the resort owners, for the tourists, for the drug dealers who need remote unpopulated areas—for whom?" Sutherland suggests the adage, "follow the money" (135). Or in other words *cui bono*, who benefits? It is definitely not the local population.

"Military rhetoric is often combined with new age religiosity in the new eco-colonialism much as it was with Christianity in the old colonialism" (139). She adds that in some cases, "there is a use of military language with the saving of the environment as the prize, and the people of the country as the enemy" (140). Military language is more than a verbal flourish or "rhetorical style." In early 1997, "the Belize Defense Force was ordered to a village ... near the Guatemalan border ... (where) they set about destroying crops, the only livelihood of the people of a Maya Ketchi village" (142).

Some contemporary movements for the environment or animal rights wish us to accept the purity of their motives and refuse to explore their own past or allow us to do so. In some cases, "dedicated conservationists have constructed what might be called 'appropriate' history—indeed a proselytizing one—ignoring considerations other than current conservation preoccupations" (Carruthers 1989b, 188). Conservation policies were often predicated on a "romanticized past" embodied in the "natural landscape and its wildlife" and in which the autochthonous inhabitants, Africans for example, were excluded (215). National parks were created as "fantasy worlds, enshrining the olden-day values of romantic nature" by which it was no longer possible for "society as a whole" to live. In addition, parks in "South Africa, as in other countries," represented "atonement" for the earlier "killing of wildlife" (Carruthers 1989a, 32; Carruthers 1995b, 176). In other words, the autochthonous inhabitants, whose past and views on the land were disregarded, were sacrificed to pay for a dominant group's fantasies and sense of guilt about their past behavior.

Wildlife and Animal Rights

Conservation and animal rights groups argue that nature or animals cannot speak for themselves. Who will speak for nature, they ask. Who speaks for the jaguar? As we noted previously, those who claim to speak for the jaguar by forming a preserve in Belize to "protect" them, do so at the expense of local humans. One author notes that this type of question is "precisely like that asked by pro-life groups in the abortion debates: Who speaks for the fetus?" (Haraway 1992, 311). Haraway asks and answers what is wrong with this type of question.

> Permanently speechless, forever requiring the services of a ventriloquist, never forcing a recall vote, in each case the object or ground of representation is the representative's fondest dream..... The effectiveness of such representation depends on distancing operations. The represented must be disengaged from the surrounding ... and relocated in the authorial domain of the representative. Indeed, the effect of this magical operation is to disempower precisely those—in our case, the pregnant women and the peoples of the forest—who are "close" to the now represented "natural" object (311–312).

Haraway adds that this operation "forever authorizes the ventriloquist" and consequently "tutelage will be eternal." Those closest to the situation are deemed to be opposing interests. "The **only** actor left is the spokesperson, the one who represents" (Haraway 1992, 312; see also Brockington and Homewood 1996, 101; Knight 1999; Hitchcock 1995, 169–172). Though the arguments may be uncomfortably similar, most "animal rights advocates" are not "pro-life" (anti-abortion) or vice versa. It is yet another case where polar ends of the spectrum are more alike than they believe themselves to be. And as this chapter makes clear, it is those who fervently claim to be against globalization and the destruction of the rights and livelihood of the world's disadvantaged who are themselves the most avid proponents of policies that strip away the rights of local inhabitants in the name of preserving habitat and species diversity, particularly involving what has become known as the "charismatic megafauna" such as elephants.

A global concern for the environment and the call to "think globally and act locally," while lofty and harmless in practice, have a tendency to become a crusade (to "think globally, impose locally") that is devoid of notions of social justice and a concern for local peoples' perceptions (Mowforth and Munt 1998, 181).

CHAPTER 3

Life in the Bush

To its credit, anthropology as a discipline led us away from chauvinism to a greater appreciation of the richness and diversity of the many ways of being human. In the excess of antitechnology fervor that has gripped the industrial nations, from the age of Aquarius to harmonic convergence, many an anthropologist has caught the postmodernist fever. We have been told of the superiority of hunting and gathering societies over their agricultural successors: The hunters and gatherers were the "original affluent society" (Sahlins 1972, Chapter 1; see also Cook 1974; Neale 1973 for a more critical view; Hill and Hurtado 1989, 436–443). "Want not, lack not," we are told, is the real basis of affluence (11). In fact, for hunters where movement is critical for success, "wealth is a burden" or an "encumbrance." Stated differently, "mobility and property are in contradiction" (11–12). In contrast to the economic man, the hunter's "wants are scarce and his means (in relation) plentiful" (13). Consequently, we shouldn't think of hunters as "poor" but as "free" (14; see also Gowdy 1998; DeGregori 1998). They have, to one author, created "the most successful lifestyle humans have yet devised" (Gowdy 1994, 27). The early hunters and gatherers (and their modern counterparts) might also be thought of as the original leisure society. Their failure to "build culture" is not, as some might believe, "from want of time. It is from idle hands" (Sahlins 1972, 20). "The evolution of economy has two contradictory movements: enriching but at the same time impoverishing, appropriating in relation to nature but expropriating in relation to man. Poverty "is an invention of civilization" (37).

As with most writers on Paleolithic hunters and gatherers, Sahlins warns about some of the dangers of using contemporary hunting and

gathering societies as a basis for understanding the lifeways of Paleolithic peoples (Sahlins 1972, 8–9). Or as another writer has stated it,
contemporary hunters and gatherers are not "fossils of the Stone Age"
(Friedman 1994, 12). Given the frequently found conditions of duress
in contemporary hunting and gathering peoples, Sahlins argues that
they are, then, not only a "fair test of the hunter's productive capaci-
ties," they are a "supreme test." But in the way that Sahlins frames the
question, it is not really a test at all. For given the duress, if their condition is found to be less than ideal, it can be attributed to external factors. Basically, as Isenberg notes, Sahlins ignores the "precariousness"
of hunting societies, relying "primarily on mammal hunting for subsistence." What is ignored is the "historical process—the interaction of
ecology, economy, and culture—that produced them." Whether it be a
recent or contemporary hunting-gathering society or one before the
emergence of agriculture, the sustainability of these societies assumes
a "timelessness; it posits the ability of human societies to arrest economic, cultural and environmental change" (Isenberg 2000, 91). If any
of the ideal (or idealized) conditions as described previously by Sahlins
are found to be true, they are almost by definition to be attributed to
their hunting and gathering lifeway. It is a test of a kind that we would
all like to take, one in which it is impossible to fail. It is one used all
too often by those like Sahlins who wish to exalt a way of life that is
impossible for virtually all of humankind.

Groups like !Kung (also known as Ju/'hoansi) Bushmen (San) have
been romanticized as simple, fun loving, harmless people in books and
intellectually racist films such as Jamie Uys's film, *The Gods Must Be
Crazy*. Kxao, the star of the film, who now lives in a brick house, states that
"films that show us in traditional clothes carrying out traditional activities
only show the past" (Jeursen 1996; see also Worsdale 1996). Tourists "see
people wearing normal clothes and to them they are not Bushmen"
(Jeursen 1996). "Their first question is always 'Where are the Bushmen?'"
(Jeursen 1996; see also, Hitchcock and Brandenburgh 1990, 22–23).
Elsewhere, Survival International describes the following scenario:

> The scene is a Bushman camp in remote Botswana. In the distance a
> plume of dust shows the arrival of a jeep. The people, whatever they are
> doing, quickly pull off their T-shirts, trousers and cotton dresses, and
> begin to dance (Survival International 1991).

Because of pressure from the conservation groups, "the Bushmen
are increasingly restricted in what they can hunt" (Armstrong 1996).

These restrictions in many parts of Africa have kept people from "utilizing wild animal resources in their own areas." This absence of "local control over natural resources" has, in one observer's judgment, contributed to the "problems of environmental degradation and decline" and not to the conservation of wild resources (Hitchcock 1993a, 132). Not only are restrictive practices harmful to more vulnerable human beings but, as we will argue, they are based upon conceptions of "nature" that have no basis in scientific fact (Drury 1998, 193, 197).

It is claimed that "because of limited access to resources and wildlife, Bushmen are caught poaching and they're tortured by wildlife officers" (Armstrong 1996). In another preserve, where the Ju'/wasi Bushmen are themselves "the main attraction," they are "allowed to hunt—but with bow and arrows only" presumably so the game species will not be threatened (Jeursen 1996; see also Davies 1998; Hitchcock and Holm 1993, 326; and for another group, the Hadza, see Hitchcock 1993a, 144).

> Lions regularly take their donkeys and goats at night ... but Bushmen may not hunt them down, even if they take a child because lions are protected game. Ostriches, too, are protected so they are no longer free to collect eggs for their delicate jewelry (Armstrong 1996).

The rights of the autochthonous population are to be protected only insofar as they act in terms of our romantic preconceptions of their "true" lifeway. Or, as is the case with the Bushmen, they are not recognized as the indigenous inhabitants of the areas in which they now live. In many places, Bushmen "found themselves reduced to the status of squatters on their ancestral hunting ground" or forced out of it (Armstrong 1996; Nicoll 1997). Under colonialism, it was the prerogative of Europeans "to determine the character of primitive culture," a prerogative that has now been assumed by contemporary environmentalists (Neumann 1998, 128).

It should come as no surprise that Bushmen want "new housing, new clothes and hunting dogs" like most everyone else. When a group of Bushmen living in the Kalahari Gemsbok National Park (South Africa) expressed an interest in such non-traditional accoutrements, the park wardens concluded that "their desirability as a tourist attraction is under serious doubt, as is the desirability of letting them stay for an indefinite period in the park" (Survival International 1995). "The 300-strong community of Kalahari Bushmen lived in what is now the Kalahari Gemsbok Park for generations until it was proclaimed a national park in 1931." The Bushmen were allowed to stay there until 1937, when they were forcibly resettled on land south of the park because "they were perceived to have acquired modern habits" (Macleod 1998). The 1930s park policy is described by one perceptive observer as "conserving the Bushman to extinction" (Gordon 1999).

There are similar arrangements for Bushmen in "Kagga Kamma Game Park north of Cape Town, where tourists can view them for $7.00 ($1.50 of the fee goes to the Bushman)" (Kirshenblatt-Gimblett 1998, 163). "The ad hoc groups who have chosen to perform as 'authentic Bushmen' are dressed in skins, while their prepubescent children go naked" (Bester and Buntman 1999, 50). A game lodge owner who sought Bushmen "to perform their ancient traditions" rejected those offered because they did not look like genuine Bushmen (Koch 1995; Bester and Buntman 1999, 94).

The Bushmen now have their own reserve near Kagga Kamma where they can compete for the tourist money. In response to a shortage of Bushmen at Kagga Kamma, authorities hired Cape Coloured to pose as Bushmen, much to the dismay of the Bushmen who did not appreciate either the cultural appropriation or the competition. "'They are taking on coloured people to carry on our Bushman traditions. I don't want my tradition tampered with like this'—so says Khomani Bushman leader Dawid Kruiper about the Kagga Kamma private game park which hired Coloured people to pose as 'Bushmen' as a tourist attraction" (Wilmsen 1999; ZA 1999).

For "eco-theme" park developers, Bushmen have been much in demand even in areas where they are not indigenous or at least have not occupied for centuries, if they ever did. The general manager for the wealthy American who before his death sought to develop a park in Mozambique, wanted to "import a group of Bushmen from the Kalahari into the Mozambican theme park" in spite of advice of "South African consultants" who argued it was "likely to discredit the project" (Koch 1996).

> If I get my way, I'll bring some of them little guys out here. Can you imagine tourists on the steam train looking out of one window and seeing elephants and rhino? Then they'll look out of the other and see the little bastards running around with their loin clothes and poison-tipped arrows ... The way I see it we'll bring them rhino here and save them from going extinct so why not bring the little guys who are also going extinct? (Koch 1996).

Under apartheid, Bushmen around Gemsbok were forced to "live off meager earnings as laborers and guides, receiving a small government pension when they grew old." Inevitably "commercial film crews looking for new twists to their African adventure movies recognized their possibilities as hunters and shamans in a modern make-believe world and moved swiftly to make a series of films" (Schrire 1995, 212–213). An agreement being worked out with the Mandela-led government allows them them to "jointly own and manage" a portion of the park, to "use traditional resources in the park" and to participate in "commercial decision-making about gate fees, rest camps and 4x4 trails" (Macleod 1998; see also Yeld 1999; Sithole 1999).

Clothing, or lack of it, has often been used by authorities as the defining characteristic of indigenous populations. For example, in the Philippines, "scanty 'tribal' attire was apparently one of the main criteria used in the 1970s by the now-defunct Presidential Assistance for National Minorities (PANAMIN) to determine who received government assistance" (Paredes 1997, 7). "Fully clothed or natives clothed in lowlander's attire were rejected and forced to remove their clothing" (ICL 1979, 6). Oona Thommes Paredes notes the tendency throughout the world to stress the "obvious aspects" of ethnic identity "to fulfill the expectations of others" and the requirement that "ethnic minorities must 'act like natives' in order to receive any positive attention" (Paredes 1997, 7; Eder 1994, 36). Defining a people by their clothing or other cultural artifacts, however appreciative the observer attempts to be, can be painful for those who are the objects of this presumed "appreciation." According to Rigoberta Menchu, a Guatemalan Quiche Indian and Nobel Peace Prize winner: "What hurts Indians most is that our costumes are considered beautiful, but it's as if the person wearing it didn't exist'" (Survival International 1995).

An exhibit of photographs in Cape Town, South Africa and an accompanying book show the changing portrait of Bushmen from evil demons to harmless people (Sharp and Douglas 1996; Skotnes 1996). Kidnapping "natives" and bringing them home for display is as old as colonial exploration. In the late 19th century, it became common to bring groups—generally by force or fraud—of scantily clad "natives" from Africa or Asia to perform or tour England or the United States.

[I]n 1905 a group of Batwa pygmies from the Ituri forest region of the Congo were brought to England ... for a season at the Hippodrome theatre along with such "entertainments" as seventeen polar bears ... jugglers and circus acts. Batwa appeared on the stage "wearing only girdles of long grass fastened around their waists and necklaces of beads", the males armed with spears, or bows and arrows, and one with a tom-tom, against a "scenic background representing their village" (Street 1992, 126–127).

Such exhibitions were portrayed as being ethnographic and scientific. The tourists now go to the "natives," and it is considered "cultural." Bringing the "natives" to Europe or North America, or bringing the tourists to the natives, are both demeaning to the peoples who serve as an attraction to sate others peoples' curiosity. In one case in South Africa, a farmer turned an unsuccessful agricultural venture into a tourist attraction where Bushmen were re-settled and could "earn a living making weapons and crafts for sale" (Schrire 1995, 213–216).

Using the indigenous population as a tourist attraction has been called the "zoofication" of the indigenous population (Mowforth and Munt 1998, 273–276). In Southern Africa, it is also called the "museumization" of the Bushman.

Tragically, under the apartheid regime in South Africa, "one of every four San" was either "in the army or an army dependent ... (which was) the highest rate of military service of any ethnic group in the world" (Gordon 1988, 17–18; see also Gordon 1992; Gordon and Douglas 2000, 1–2; Thurow 1989). They were ruthlessly exploited by the white minority government in Pretoria (Uys 1998). Given the role they were forced to play, their future in independent, majority-ruled Namibia is not much brighter. In a recent conflict over land, Kipi George, "the elected chief of the Kxoe," argued:

> some people within the government are still trying to punish the Kxoe for having taken the wrong side in the Namibian liberation struggle. Between 1975 and 1989 the South African Army used attractive wages and racial propaganda to persuade thousands of "bushmen" soldiers to serve as trackers and reconnaissance troops (O'Loughlin 1997).

Kipi George adds that some people say to them: "We remember you when you were killing us." Kipi George adds: "Every tribal group in Namibia has members who fought against [the government], but we are the only ones who are being blamed" (O'Loughlin 1997). The white leaders of their military units are becoming wealthy while the Bushmen former fighters live in squalor in refugee camps. Some Bushmen groups have already found it necessary to trek north to Angola where the conditions are far from ideal (Inambao 1997).

One author, Edwin Wilmsen, argues that the Bushmen are not ves-
tigial hunters and gatherers but a marginalized, impoverished people
who lost economic and political power and status as a result of a series
of major economic transformations in Southern Africa in the 19th century (Wilmsen 1989, 157; see also the excellent review of Wilmsen
1989 in Gordon 1990, 18–19; Wilmsen and Denbow 1991; Schrire
1980; Hitchcock and Holm 1993). Rather than living in an environment with abundant food resources, the Bushmen experience periods
of severe food deprivation and are generally malnourished. Wilmsen
concedes that Bushman morphology is not "entirely environmentally
determined," but he argues forcefully (with considerable weight,
height, and nutrition evidence from other researchers) that a large portion of the differences in weight can be explained by relative degrees
of impoverishment (Wilmsen 1989, 303–315). "False assumptions
force the equally false view that any qualitatively original introduction
into that assumed static 'forager formation' (which ... has not been a
fully independent social formation for more than a millennium)
requires change in its processes of goal attainment, integration, and
adaptation" (317; see also Wilmsen 1990; Wilmsen 1983). Stated differently, their assumptions about a Bushman stasis with their environment lead to policy prescriptions that at best minimize the pace of
change and help to perpetuate their condition of poverty.

Konner and Shostak, who admit to being participants in the creation
of the romantic myths about Bushmen, suggest that we project on other
cultures and peoples (or earlier times than our own) a vision of life as
we would like to it be. What we find missing in our lives we miracu-
lously find in others'. This is often the case when there is a desire to
prove a proposition about the basic human condition. What Konner and
Shostak show is that, to sustain this vision, they and others overlooked
data in their own and others' outstanding field work, data that almost
leaps off the page in their representation of it (Konner and Shostak
1986; see also Howell 1988; Howell 1976). The Caldwells similarly
argue that those who support the thesis of "primitive affluence," using
the descriptions of earlier travelers to a region, overlook other passages
where the same authors describe hunger and high infant death rates
(Caldwell et al. 1987, 31–32).

In an article in a book on the Tasaday, Richard Lee makes an argument much like that of Konner and Shostak about romanticizing the
Bushman as exemplified by films such as *The Gods Must Be Crazy* and
The Gods Must be Crazy, Part 2. "Here we have the same kind of misrepresentation of the foragers with a timeless and beautiful vision of

simplicity and natural harmony crowding out the far less pleasant realities of life among former foragers." Further, they "take the raw material of the primitive 'Other' and seek to construct hegemonic ideologies
that mirror the temper of the times and reinforce prevailing values ...
ideologies centering on the loss of innocence and shattered illusions."
That perceptive passage by Richard Lee would find almost complete
agreement from "revisionists" except for the designation of Bushmen
as being "former foragers" (Lee 1992, 170–171). The irony is that Lee
has been the main participant in the debate with the revisionists on the
contemporary condition of Bushman society.

Gordon asks the question why "almost all contemporary anthropology texts *still portray* Bushman as if they live in a state of 'primitive
affluence.' There is no simple answer to that question" (Gordon 1992,
2000, 3). He refers to Nancy Howell, "one of the most prominent members of the Harvard Kalahari Project" who admits to ignoring "the para-
phernalia of Western civilization and poverty ... because we didn't
come all the way around the world to see them. We could have stayed
at home and seen people behaving as rural proletariat" but only the
"Kalahari and a few other remote locations" allowed us a "glimpse of
the 'hunting and gathering way of life'"(Howell 1988; see also Howell
1976).

> So we focus upon bush camps, upon hunting, upon old fashioned cus-
> toms, and although we remind each other once in a while not to be
> romantic, we consciously and unconsciously neglect and avoid the
> !Kung who don't conform to our expectations (Howell 1988).

Not only have we projected our desires as behavioral traits of other
peoples, but researchers have also made similar projections in descriptions of other primates and have later had to correct these erroneous
interpretations (Konner and Shostak 1986, 74–75). Possibly the idyllic
life of the !Kung Bushmen observed by Lee and others may yet have
been another role or performance played by them, unwittingly bought
and paid for by the anthropologist-observer. In a now famous essay,
Lee admits to having provided the !Kung Bushmen with tobacco, medicine, metal tools, and a Christmas dinner. He was the "only source of
tobacco in a thousand square miles" and he would at times cut off individuals for a few days because of "non-cooperation" (Lee 1969). An
often-cited 1948 study for the primitive affluence thesis of the aborigines of Arnhem Land in Australia also noted "just how artificial the way
of life had become because of food available from the mission stations,

which had made hunting and gathering optional." The Caldwells add that even so, the researchers calculated "an infant mortality rate of 127 per thousand" which the Caldwells considered to be "undoubtedly much below the true rate" because of problems in data collection (Caldwell 1987, 32).

The issue is not whether the Bushman or other groups still retained the skills to carry on with the "traditional life" during the time of the anthropologist's visit, but whether they would have done so or could do so sustainably over a longer time frame without outside contribution. There is, to say the least, a difference between a lifestyle paid for by outside observers for the purpose of observation and one that is freely chosen from among other options. Critics of the Harvard Kalahari project have noted the exclusion of "trained" ethnohistorians from the project, in spite of the fact that "innumerable archival sources existed in places such as Cape Town, Gaborone, Windhoek and London" (Schrire 1984b, 11; Gordon 1984).

It should be noted before we proceed further, that however erroneous were these romantic descriptions of Bushman life, they were a great improvement over the previous colonial studies of the "savages" and laid the empirical foundation upon which the corrective perspectives are established. Many of the pre-romantic views still persist and support policies even more detrimental to the Bushman than those we are critiquing (Gordon 1986). In an analysis similar to that of Wilmsen, Sumit Guha demonstrates a clear linkage between colonial racialist theories, the "new environmental consciousness," and other modern romantic views of tribal peoples in India (S. Guha 1998, 432–433). Guha cites an environmental report about tribal people "who from time immemorial have lived in total harmony with forests" (CSE 1986, 376). Guha adds that "we see again the picture of the timeless harmony with nature disturbed only in recent times by the intrusive forces of the state," a picture that Guha shows is fundamentally at variance with historical fact (Guha 1998, 432; see also Fox 1969).

Neumann refers to the "mythical vision of Africa as an unspoiled wilderness, where nature existed undisturbed by destructive human intervention." This "European conception of unspoiled nature" was part and parcel of the European "concept of primitive human society" all of which "had more to do with European myths and desires than with reality" (Neumann 1998, 128). In other words, "Eden had been rediscovered" (Neumann 1998, 32).

In other parts of Africa, the "Garden of Eden" of abundant wildlife and sparse population was often the result of crashes of human and livestock populations as a result of colonial conquest and diseases that had recently been introduced (Bell 1987, 8–9).

Finding some inherent superiority in other races or cultures or levels of technological achievement is as silly and probably as dangerous as finding others to be inferior. Lionel Tiger is an anthropologist and author who has written extensively on evolution and its implications for understanding human behavior. To Tiger, the transition to agriculture was not made because humans wanted to farm but because of population pressures (Tiger 1987, 35–36; see also Cohen 1977). The behavioral differences between the genders became wider, the community became too large and fragmented, and a "moral missing link" between humans was created. Tiger sees many phenomena of modern society, such as the nightly news, as a vain attempt to re-create the integrity we once had in small groups (Tiger 1987, 17–69).

Mark Nathan Cohen is frequently cited on the thesis that the transition to agriculture was the result of resource stress, as the population had reached the limits of environmental support by hunting and gathering (Cohen 1977). If this resource stress occurred when there were at most 5 to 10 million humans in the world, how, pray tell, can a global population of 6 billion return to a lifeway of foraging? The lives of the pre-neolithic populations were short and so were those of the early agricultural societies—*probably* ranging from the low to high 20s in life expectancy. Life expectancies reached "30 years with consistency only under civilization" (Howell 1976, 35). We have come a long way since.

What is too often not asked by those who feel conditions were worse after the transition to agriculture is what would have happened to the hunting and gathering peoples had the transition to agriculture not been made. In other words, a before and after comparison is invalid; one must compare the options and their consequences at a particular time. However extraordinarily slow population growth was, it was positive. This means that "resource stress" would have continued to worsen and a sustainable human process would have been possible only by decreasing population growth. Given the history of humans until then, the likely adjustment, absent technological change, would have been increasing the already very high death rates. Whatever the negatives of the agricultural revolution were, it did put humanity on a path to what has until now been a sustainable expansion in human life. If humans were forced by "resource stress" to make the transition to agriculture because of population growth, however slow it may have been, it means that the hunter-gatherer regime was not "sustainable" in any meaningful sense of that term.

Tiger and others eulogize the !Kung (Ju/'hoansi) for their sharing, lack of stress, and avoidance of hostility. Lorna Marshall, who is the mother of Elizabeth Marshall Thomas, author of the famed book about !Kung, *The Harmless People* (Thomas 1959), said of the !Kung that she did not have the "fortitude to learn more" about how the women disposed of their babies out of "necessity" because of "the meagerness of the resource of food and water" (Marshall 1960, 327–329; on infanticide for similar reasons among the Saniyo-Hiyowe of Papua New Guinea, see McElroy and Townsend 1989, 130–135; on evidence of cannibalism in New Guinea and earlier in Puebloan culture, see Diamond 2000; Marlar et al. 2000). If such practices shock readers of Western cultural heritage, as they should, then it must be noted that in earlier, poorer periods in Western culture there was considerable infanticide and abandonment of children (See Boswell 1988; DeGregori 1985a, 184–185, and 217–218 for additional bibliography). Sahlins admits to the existence of "infanticide, senilicide, sexual continence for the duration of the nursing period, etc." among hunters and gatherers. Sahlins lumps these together as seemingly of comparable importance in population control among hunters and gatherers, attributing this not to lack of food resources but to the need for mobility (Sahlins 1972, 34). Even Lee, whose views of the Bushman are in line with Sahlins on hunter-gatherer life, speaks of "occasional infanticide" which, along with high infant mortality rates, kept the population in check (Lee 1972, 337).

Tiger does not give us Marshall's observations on the practice of infanticide or the high infant and child mortality rates (40 percent die before age 15) as noted by Richard Lee; !Kung homicide rates calculated by Lee were higher than those of New York City. And there were rampant and prevalent infectious diseases such as influenza, pneumonia, bronchitis, gastroenteritis, rheumatic fever, tuberculosis, and gonorrhea, as observed by Marjorie Shostak, who also noted that now, with increased contact with the outside world, the !Kung were "actively seeking" possibilities for change. Increased contact with the outside world means that these indigenous peoples, "if they are to survive, must gain access to resources, including land, labor, tools, and capital" as well as "land tenure security" (Hitchcock 1993, 145, 149). Among other things, the Bushmen wish to own cattle like their Tswana neighbors (Hitchcock 1993, 136). Fortunately, the Bushmen and others with a recent tradition of foraging cannot read Tiger, who presumably thinks they shouldn't be able to read (Shostak 1981, 15, 110, 201, 349; Konner 1987, 8–10; Knauf 1987, 46, 48; on homicide rates among the

Bushmen and other "gentle people" see Bower 1988, 90–91; Chagnon 1988, 985–992; see Lee and DeVore 1976; Lee and Solway and Lee 1990, 109–146 for a perspective more in line with Tiger and Sahlins).

CHAPTER 4

Paradise in the Pacific?

The myth of a pre-European contact paradise has been a staple of our perceptions of the island civilizations of the Pacific. In fact, for many, this Edenic perception has persisted to the present. The reality is considerably different. Famine has been a recurrent phenomenon in the Pacific Islands both pre- and post-European contact (Kirch 1984, 128–131). "There certainly was deforestation, famine, warfare, collapse of civilization and population decline" (Bahn and Flenley 1992, 212). "Hawaii which is often seen as a Pacific paradise ... has a recurrence interval for famines during the 190 years since contact of only 21 years. Such a recurrence is no less frequent than the historical record of famines in Bangladesh" (Currey 1980, 447–448).

Bahn and Flenley use Easter Island as a metaphor for what could happen to the entire planet. Like our planet, "Easter Island was an isolated system" with the people believing "that they were the only survivors on Earth, all other land having sunk beneath the sea" (Bahn and Flenley 1992, 212–213). Bahn and Flenley found that the Easter Island inhabitants permitted:

> unrestricted population growth, profligate use of resources, destruction of the environment, and boundless confidence in their religion to take care of the future. The result was an ecological disaster leading to a population crash (Bahn and Flenley 1992, 212–213).

Bahn and Flenley argue that a "crash on a similar scale (60 percent reduction) for the planet Earth would lead to deaths of about 1.8 billion people, roughly 100 times the death toll of the Second World War"

(Bahn and Flenley 1992, 212–213). If they are correct in their percentage decline (60 percent), the comparable global scale would be upward
of 3.6 billion deaths (for the world population in 2000) or double their
estimate for the early 1990s.

> When Polynesians settled Easter around A.D. 400, the island was cov-
> ered by forest that they gradually proceeded to clear, in order to plant
> gardens and to obtain logs for canoes and for erecting statues. By around
> 1500 the human population was about 7000 (over 150 per square mile),
> about 1,000 statues had been carved, and at least 324 of those statues
> had been erected. But—the forest had been destroyed so thoroughly that
> not a single tree survived (Diamond 1994, 51).

I have checked with regional specialists who have verified that not
"*a single tree survived.*" Yet the myth persists that "prehistoric Oceanic
peoples avidly practiced a 'conservation ethic' toward their island habitat" (Kirch 1984, 123). One anthropologist has referred to our imagining that the "pre-European Pacific was a paradise of holistic healing,
ecological reverence, love for the land, and communalism" (Keesing
1990, 169; for a counterview, see Trask 1991; on traditions in the
Pacific, see Fry 1997; J. Turner 1997). They were seen to be "actors on
a changeless stage" (Kirch 1997, 4).

"Recent evidence shows this view to be false, and one suspects that the
true scale of prehistoric impact on the Pacific Islands is not yet
fully grasped either by prehistorians or natural scientists" (Kirch 1984,
123; see also Kirch 1982; Kirch and Lepofsky 1993; Kirch and Hunt
1997; Bahn and Flenley 1992; Athens and Ward 1993). It is interesting
that the 18th century European visitors to the islands such as Louis de
Bougainville and James Cook found them to be an Edenic paradise in
line with the romantic notions of the period of "noble savages" and
homme naturel (Kirch 1997a, 4–5; Spriggs 1997, 101). However benign
this view may seem, it did not prevent the population of the islands from
being "dispossessed of their land" (Spriggs 1997, 101–102).

The islanders did what people elsewhere did; they hunted many animals into extinction, poisoned fish, burnt grassland, and deforested
large areas, leading to soil erosion (Johannes 1978; Clarke 1971;
Clarke 1990, 235–242; Kirch 1984, 137–148; for a history of Western
romantic views of the Pacific Islands, see Adams 1983, 234–237; on
bird extinctions, see Kirch 1997a, 4, 11; Steadman 1995, 1997). In New
Zealand, they began "mining the easily accessible and protein-rich
seam of big-game resources" until these "game sources declined sig-

nificantly." This hunting practice is best described as an "economic-optimization model" (Anderson 1997, 283). There were also extinctions of "certain species of giant marsupials" in Australia and Tasmania where the "arrival of aborigines and their weapons and fire-sticks were possibly part of the web of death" (Blainey 1976, 58; see also Flannery 1999; Diamond 1997, 42–47 for very readable, scholarly, but also popular accounts).

Anderson refers to the "radical New Age assertions that Maori subsistence behavior was actuated by a deep sense of ecological relationships coupled with a mystical reverence for the environment." He adds that the "evidential basis for these assertions is highly dubious in all respects (Anderson 1997, 273).

Spriggs has found that "Pacific nations people find it easy to live with the idea" that they and their ancestors have "actively altered their island environments." It is in the "terminally colonized Pacific countries" where the later European immigrant population is dominant that people "often talk of a golden age before the arrival of the 'white man' when their ancestors lived in harmony with the environment" (Spriggs 1997, 101). However much New Agers may pride themselves on their own enlightenment and multiculturalism, here as elsewhere it is in reality a further exploitation of a conquered people who are a minority in their own ancestral land. These people have legitimate grievances that need to be addressed and will only be addressed by a combination of intelligent, militant activism and a positive response from at least some members of the dominant community. New Age romanticism and mysticism are merely a retreat into obscurantism; expressing grievances in these terms makes proponents seem ridiculous in the eyes of those who might otherwise be sympathetic to their cause.

For Easter Island, Bahn and Flenley admit that there could have been a "major drought," but then they add that "it seems odd that the forest should survive for at least 37,000 years, including the major climatic fluctuations of the last ice age and postglacial climatic peak, only to succumb to drought after people arrived on the island" (Bahn and Flenley 1992, 212).

Millions of years ago, a group of islands arose in the Pacific and because of continental drift passed between what was then the gap between the North and South American continents, forming part of the chain of Caribbean Islands. This drift occurred long before the emergence of *Homo sapiens*, so there is no human historical connection. However, in the Western imagination and the environmentalist rhetoric, there is the shared identity of a pre-European contact paradise or at least an environmentally benign habitation. This is, as we previously argue, definitely not the case for the Pacific Islands nor is it the case for the Caribbean. A study published in *Science* in July 2001 refutes any such contention (Jackson et al. 2001). As one of the authors, Karen Bjordal, a zoology professor at the University of Florida states: "There's been a longtime belief that everything was fine until the ... Europeans showed up.... Now we've discovered that the start of the environmental problems (in the sea) go way back before that" (Recer 2001). Another of the authors, Charles Peterson of the University of North Carolina at Chapel Hill added: "The notion of the native peoples having a benign impact on the environment in their vicinity has been challenged.... The general feeling is that there were dramatic effects locally and not a prudent predation" by ancient humans long before the Colonial and industrial eras (Recer 2001).

"There are dozens of places in the Caribbean named after large sea turtles whose adult populations now number in the tens of thousands rather than the tens of millions of a few centuries ago." Prior to the 19th century, "vast populations of very large green turtles were eliminated from the Americas (Jackson et al. 2001). "Algae now choking and killing many coral reefs in the Caribbean can be traced to the slaughter more than 3,000 years ago of the green sea turtle and to other animals that grazed on the sea plant." The kitchen refuse piles of the first American Indian settlers in the Caribbean indicate a heavy reliance on the sea turtle for food. "The animals were easy to catch as they regularly lumbered ashore to lay eggs on the semitropical islands." As the turtles' numbers diminished with the slaughter, so too did the kitchen refuse diminish through time until "the turtles disappeared entirely. It is clear the nesting colonies were wiped out," Bjordal said. As the turtle population was decimated, other fish were harvested, "such as the large parrot fish, a meaty dweller of the reef. Those, too, eventually became scarce, as did other plant-eating animals" (Recer 2001).

Many still view these island states as mini-utopias, even though there are economic problems causing major migration of populations. In small countries and islands, such as those in the South Pacific and the Caribbean, lack of opportunity (employment or otherwise) leads to migration from rural areas to towns, from smaller islands to larger ones, and from the islands to more developed areas such as Australia, New Zealand or the United States. How are you going to keep them down on the family plot in Samoa after they've seen L.A. (or Sydney or Auckland)? These cities will frequently have more of an island's population than the island itself. For developed country lovers of Gauguin or

Maugham, the tropical islands may seem like paradise, but not to those who must earn a living there. Those most likely to migrate are the young able-bodied, and the better educated (the brain drain), leaving behind a population that is disproportionately old and very young. The migrant remittances in many cases are sustaining the economy, providing 30 percent or more of its income. Labor becomes the island/countries' most important export (Poirine 1998; see also R. Brown 1998; Poirine gives both the detriments and benefits of "investment" in labor for export). In many if not most instances (for Pacific Islands), the combination of migrant remittances and external aid provides well in excess of 50 percent of the income of the island. For one Pacific island, aid has been 70 percent of the gross domestic product. Yet most of the islands of the South Pacific are deeply in debt (AP 1997). In Western Samoa, remittances and aid grants totaled more than four times the domestic exports in 1989 (Ward 1993, 9). In some places that have experienced recent migration of the young and an increase in life expectancy, there are problems of both population growth and aging without the high levels of development that normally cause them.

For the migrant workers like those from the Pacific Islands, the new information technology has added another magnificent dimension to the cultural contact that followed advances in air travel and communications from the late 1960s onward. Those with access to the Internet can receive daily email news bulletins from "home," provided free by the embassies of many countries. Or they can use the World Wide Web to access local newspapers and other news sources from their former homeland. Email gives a sense of immediacy to "letter writing." One of my students and her husband are able to use the Internet to access the television station of their small (population less than 100,000 and falling) eastern Caribbean island.

Even small and far off islands in the Pacific with a population of a few thousand (with more abroad than home) have home pages and bulletin boards so that expatriates can communicate and retain a sense of community with the home island and other expatriates dispersed around the world. The Pacific island of Rotuma has a population estimated at 12,000, but only about 2,600 actually live on the island. It has a Web site that was visited over 15,000 times in its first year of operation, even though cost factors make it prohibitively expensive *at present* for those on the island to access it (Howard, 1999, 162, 172). For the Pacific island region, there is now a Web site—Kava Bowl: The Pacific Forum—which can be used to access discussion forums "for a number of island-centered communities" (Howard, 1999, 162; Ogden, 1999, 461). There are many islands where high fertility rates and limited economic opportunities have meant rapid population growth and a rapid out-migration of population such as the migration from the island of Rotuma. This means that the majority of the people of these island heritages were off the island and in some cases scattered around the world.

The viability and sustainability of these communities could once be said to be in doubt. Since the proverbial dawn of history, small groups have been assimilated into larger groups, losing much of their culture and identity. Modern states and modern communications have seemed to many to have accelerated this process. Quite possibly, the "virtual" community in cyberspace might be the binding force that unites a people and keeps a community and its traditions alive and sustainable even as they live and participate economically in larger communities. It is by no means a certainty, but the possibility and hope is there.

Obviously all newly arrived migrants do not have immediate access to all of the information technologies. But there is far more available, and the entry level is lower than it has ever been, making the various technologies more accessible to more people than ever before. Most readers would probably be pleasantly surprised to learn how many are able to use this technology in order to maintain contact with distant loved ones and with a cultural heritage. And it will be continued technological change that will create new technologies of contact and heritage maintenance, lower the cost of existing technologies, and give the possibility of advances in income that allow an increasing number of people the opportunity to use these technologies. These same technologies allow others greater opportunity to become informed about a diversity of peoples and cultures.

The Pacific paradise of old is a product of a new colonial mind that operates in an intellectual wilderness. True, the new mind has altered the value structure to pay homage to what was previously condemned or looked upon with a paternalistic sense of the superiority of the conquerors' culture. But those who reprint these alleged ecological statements or propagate myths of harmonious lifestyles of Pacific Islanders or American Indians do not go to the trouble of authenticating them, since they already find in them an echo of what they wish to see in themselves. In a word, they are fables, and it is best to keep reality as a Terra Incognito. Thus the romantic living in our scientifically and technologically advanced society finds in other people's lives the answers that complete questions they were asking about their own lives and identities (Berry 1982, 120; Bishop 1989, 250). We can find this response only to the extent that we are ignorant of the reality of the other places. This romantic aesthetic can only "occur in empty spaces" (Berry 1982, 115). The "wilderness lies beyond Eden's walls. It is where man must go when he is cast out of paradise." It is a place "beyond human control" (Adams and McShane 1996, 6).

We all respect those who think well of other people, groups, religions, nations, cultures, or even of other species. Certainly it is admirable to think well of others until proved otherwise. If romanticizing other cultures remains strictly at the level of feelings, then few of us would object. Unfortunately, much of this romanticizing arises out of philosophies, such as postmodernism, that reduce all thought to subjective feelings without taking their consequences into consideration. But, as we have shown time and again throughout this book, our ideas have consequences and must be judged in terms of them. We have tried to show that romantic falsifications are most often harmful to those who are romanticized. Bali is an excellent example of this. Few peoples have been romanticized to the extent that generations of colonial administrators, followed by anthropologists and others, ignored massive evidence to the contrary to praise the Balinese for their mysticism, pacifism, and harmonious existence. What is often astounding is the truly massive amount of contrary evidence that romanticizers have simply ignored or dismissed as being atypical. But as with other myths, the Bali myth has not been benign. As Robinson argues, "the Bali myth has helped to falsify history in a way that has served the people in power while silencing those who have suffered injustice" (Robinson 1995, 307). When a scholar such as Robinson exposes the myth in *The Dark Side of Paradise*, it is amazing how overwhelming the data are in support of his thesis, and one wonders how they could have been ignored for so long.

The late Roger Keesing cogently delineates the dangers of mystifying the history of other peoples, however worthy and noble the intent may be. He raises concerns about the "disciplinary inclinations" of anthropology towards a "liberal angst and fuzzily romantic relativism" which "has been compounded with the rhetoric of postmodernism to the point of self-mystification and auto-paralysis." Keesing finds the "ideological mythicisation of history" to be "part of the political process everywhere, in North America and Europe as well as in the Pacific." There is an "urgent priority" for a "critical analysis of this mythicisation and mystification" (Keesing 1993, 587).

In the last hundred years, mythic cultural nationalist histories glorifying ancestral pasts and cultural homelands have been deployed to kill, dis-possess, and subjugate: used to rationalize genocide, "ethnic cleansing," persecution and usurpation (Keesing 1993, 587).

Keesing continues with reference to anthropological inquiry:

To marshall post-modernist relativising arguments in defence of cultural nationalist causes we happen to believe are just, valorizing mythic his-tories as equally valid alternative narratives of the past, is to abandon whatever firm ground we might stand on and whatever principled stances we might adopt in regard to truth (Keesing 1993, 587).

Keesing, in an argument valid for all of academia, maintains that anthropologists have to reestablish their "credibility to speak out" against "myths of racial supremacy or genocidal nationalist narratives of blood, homeland and purity."

Truth, however elusive, is too precious a quality to be so easily and cheaply abandoned by academics, entrusted by their societies with unique opportunities and responsibilities to seek after it as evenhandedly as they can (Keesing 1993, 587).

The dying of indigenous or aboriginal populations serves the pur-poses of those who have an ideological agenda. Often this has necessi-tated falsification of history by denying that there were any survivors of a brutally massacred peoples, such as the Tasmanian Aboriginals. Defense of the lives and lifeways of those who have suffered domina-tion is a worthy endeavor to which people of good will should sub-scribe, but only if the defense is accurate, not an excuse to further one's own agenda. "Aboriginality can thus be cherished only insofar as it is a stable form that can be made to correspond with New Age meta-physics; Aboriginal history contributes to the picture only by showing that the relation of white culture to Aboriginal life was purely destruc-tive" (Thomas 1994, 177). In Australia where the Aboriginal popula-tion is a disadvantaged minority (as they are almost everywhere), or even in an independent island country, romantic interpretations of an indigenous culture will "almost inevitably privilege particular factions of the indigenous population who correspond best with whatever is ide-alized" (189).

The truly dark side of romanticizing other cultures is compounded when a minority of the intellectual elite of a culture or country, such as India, borrow an alien ideology such as postmodernism, and use it to defend what they consider to be their "traditional" culture against alien ideas and practices. Clearly, some alien ideas are unquestioned and privileged, while others are inherently dangerous and rejected. When there was an "incidence of widow immolation (*sati*)" in India which caused considerable soul searching among the peoples of India, one postmodernist "patriot" condemned the critics of widow immolation, "branding them as modernized Western elites who denigrate authentic folk practices" (Nanda 1998, 292, citing Nandy 1988). Meera Nanda adds, "Not surprisingly, such nativist, antimodernist ideas have found a sympathetic audience among right-wing Hindu fundamentalist parties" (Nanda 1998, 292). Conversely, the "vulgarized pop-postmodernist," Vandana Shiva, who is lionized by Western feminists and luddite postmodernist intellectuals, argues that women have an inherent "embeddedness in nature" for subsistence agriculture which would keep them in poverty and not the more highly productive modern agricultural technologies (Nanda 1998, 291; Nanda 1997, 365). It is doubtful that the Western feminists who pay homage to Shiva have thought through the implications of women having an inherent "embeddedness in nature." Carried to its logical absurdity, it would mean that the many professions and occupations that were once closed to women should have remained so, leaving women to do peasant agriculture and/or stay home and be nurturing.

CHAPTER 5

The American Indian: The "Original Ecologist"?

Finding primitive nobility in other peoples has an ironic twist. It normally does not occur until after they have died out or been militarily defeated and otherwise subjugated and therefore rendered non-threatening. In the United States, the American Indian became the subject of collections of Indian oratory of great leaders, many of whom did not know any English, which was the language of the elegant prose of these collections (See Turner, 1977; for an example of the romantic attribution, see Bahro 1986, 159). The anonymous Indian or the stereotyped Indian with a name is rendered harmless in our imaginations as long as he is kept "traditional." A chief on horseback with headdress and small traditional weapon (at most) is noble; an Indian on horseback with a carbine and several bandoliers of ammunition was a threat to conquerors and a nightmare to the romantic purists; he was a "renegade," no matter how many followers he had.

When a people have experienced the humiliation of conquest and attempted degradation of their culture, romantic falsification of their culture may be appealing to some members of the group. It is a tragic irony to learn and acquire your own alleged traditional beliefs from members of the very group that sought to destroy them. Even where the falsification becomes apparent, there are those who wish to continue to promote the myth as an embodiment of a higher truth. There is also concern that "whites will transform Indian culture in their own image." George Tinker, an Osage and a professor at the Iliff School of Theology in Denver, is worried about the "danger" that these "mutations of spirituality will make their way back into the Indian world"

(Johnston 1993). Whether it be American Indian beliefs or Buddhism or any other non-Western beliefs, romantic Westerners tend to misunderstand the belief system and will in actuality frame it in terms of their own. The New Agers have taken tribal or community-centered beliefs and practices and "centered" them on the "self, a sort of Western individualism run amok" (Tinker quoted in Johnston 1993). In India, these practices are sometimes called Karma Kola while James Eagle Bull, Lakota tribal member, deemed the pseudo practitioners to be "plastic medicine men" (Michel 1995; see also Kehoe 1990).

> Spokespeople for the environmental movement have transformed Indian respect for land and communitarianism into a cult like vision of new-age "spiritualist." ... The modern poverty of the First Nations has little to do with white suburban antagonism to industrialization or with the cult of Aquarius (Harris-Jones 1993, 49).

The contrast between the romanticized vision of American Indian life and the poverty and other problems of daily life for modern day American Indians is illustrated by the following description and dialogue from the video "Reservation Blues" created by Sherman Alexie, an American Indian. Two white "groupies" arrive on the reservation "ready to go native for the weekend."

> "You have all the wonderful things that we don't have," says one of the women. "You live at peace with the earth. You are so wise."

> An Indian member of the band responds: "You have never spent a few hours in the Powwow Tavern. I'll show you wise and peaceful." (Egan 1998, 18–19; see also Alexie 1995, 41–44).

In another encounter, a Choctaw Indian who liked "being Choctaw" and did not want "to be you" added:

> Just because I don't want to be a white man doesn't mean I want to be some mystical Indian either. Just a real human being (White 1990, 112–113; see also Krech 1999b, 59).

Harris-Jones argues that "a solution to Indian poverty" cannot be won "through a joint, heroic stand against the consumer-industrial society as consumers might wish." On the contrary, he argues that it "can only be won through sustained political support for land claims over a large area which would enable the First Nations to proceed with self-government" (Harris-Jones 1993, 49). One may or may not accept the Harris-Jones solution to the poverty problems of American Indians, but he is right on target in his critique of the New Agers. Ironically, modern gambling casinos and, to a lesser extent, royalties from mineral rights for modern mining have played an important role in lifting many American Indians out of poverty. Neither of these activities fit with the New Age category of Indian spirituality.

Romanticizing American Indians or any other group, particularly those with legitimate grievances of one kind or another, is in many respects a form of exploitation comparable to those that they have already suffered. The fictions that the romantics create about other cultures legitimizes their own beliefs and provides them with emotional sustenance they are unable to find in their own culture or civilization. It is the most marginalized, oppressed, and desperate people in other cultures that are most susceptible to this fictionalized narrative. The repeating of this mythology by marginalized peoples may warm the heart of the New Age romantics, but it undermines efforts to get their grievances seriously considered and addressed. Validation of our beliefs about others acquires a higher priority than their legitimate concerns.

Where the romanticized lifeways are of living people this external pseudo-veneration does nothing to further the culture of its members. And to the extent that the policy prescriptions of the romantics are followed, the economic advancement of these cultures can be seriously retarded. The question of falsification of other peoples' belief systems, and acceptance of this falsification by some in the group itself, goes to the heart of a much larger issue in intellectual life today. For some romantics and post-modernists, truth either doesn't matter or it is simply impossible to discern. Consequently, everyone's personal perceptions of truth are equally valid, since there are no criteria to differentiate between perspectives. The logical consequences of this argument tend to be applied to all other beliefs except one's own. Truth no longer matters, but higher spiritual conceptions and similar understandings must be accepted and cannot be subjected to any attempt at verification or falsification. As the psychologist Carl Rogers said, in reference to the writing of Carlos Casteneda—"He may be lying, but what he says is true" (de Mille 1990, 228). Concerns for verification are simply dismissed. If in fact the alleged wisdom and beliefs of a Black Elk are more those of John Neihardt, the poet who published *Black Elk Speaks*, that is of no importance. Vine Deloria, Jr. asks, "Can it matter?" He answers:

The very nature of great religious teachings is that they encompass everyone who understands them and personalities become indistinguishable from the Transcendent truth that is expressed. So let it be with *Black Elk Speaks* (Deloria 1979, XIV).

Wilderness and the Indigenous Population

As we have noted in regard to Africa, the creation of parks, wilderness areas, nature preserves and such involved forcibly displacing the existing inhabitants. In the United States:

> The movement to set aside national parks and wilderness areas followed hard on the heels of the final Indian wars, in which the prior inhabitants of these regions were rounded up and moved into reservations so that tourists could safely enjoy the illusion that they were seeing their nation in its pristine, original state—in the morning of God's own creation (Cronon 1995a, 42).

Until the last few decades, white Americans had an ambivalence about Indians and the "natural" environment. The "wilderness also represented the depraved conditions from which savages needed uplifting." Rather than being praised for having preserved the environment, they—who were variously "filthy," "dirty" and "lazy"—"only detracted from the sublimity of the scenery," making it best for all if the Indians would simply leave (Spence 1996, 42). In the United States, the situation was similar to that of the African game parks, namely "the original inhabitants were kept out by dint of force, their earlier land uses of the land redefined as inappropriate or even illegal. The Blackfeet continue to be accused of 'poaching' on the lands ... that originally belonged to them and that were ceded by treaty only with the proviso that they be permitted to hunt there" (Cronon 1995a, 42). Blackfeet were museum artifacts, "past-tense" Indians in Glacier National Park promotions and entertainment for the tourist "experience" (Spence 1996, 45).

It has been frequently and incorrectly argued that the parks were incorporating lands that were "worthless" except for their natural beauty and had been largely unoccupied by Indians (Spence 1996, 39). Yosemite was the only National Park where there remained a native community within it that initially refused to "vanish"(27). As the park developed, the Indian community borrowed cultural items and practices from Anglo and Mexican communities and from other Indian groups (31). The park visitors perceived them to be "real" Indians still living in the "exotic naturalness" of their "natural" environment doing "traditional Indian things," the marketability of which was quickly recognized by the Indians (34, 36–37).

Authenticity for the Indians of Yosemite National Park was strictly in the eyes of the park administrators and tourist beholders. For organizers of Indian Field Days, the Indians were expected to "confirm popular white conceptions of how Indians were supposed to look and behave" by conforming "to a generic representation of Plains Indian culture," including "buckskin dress, moccasin and head decoration," that was foreign to their California Indian heritage (47). When the Indians no longer served the purposes of the park administrators, they were cut off from economic participation in the park and they eventually moved to towns on the park's periphery. Yosemite National Park, like all the other national parks, would now look like an Ansel Adams photograph and fit the "image of a priori wilderness, an empty, uninhabited, primordial landscape that has been preserved as God first intended it to be." Like most myths, it served the purposes of those who believed in and promulgated it. The "wilderness," the national park dispossessed of its original inhabitants, "reaffirms the myth that North America was once a 'virgin' continent waiting to be peopled" (58).

"Nature" is more "appreciated" by refugees from urban areas who earn their livelihood from writing or from some high tech profession. "Ever since the 19th century, celebrating wilderness has been an activity mainly for well-to-do city folks. Country people generally know far too much about working the land to regard unworked land as their ideal" (Cronon 1995a). What some of us grew up believing was "wilderness" was, like most of the rest of the world, anthropogenically transformed in the past. The Gila Primitive Area or Gila Wilderness in western New Mexico was considered by many to be the last wilderness area in the United States. Or at least it was repeatedly claimed to be that when I grew up in New Mexico. "In reality, of course, the land was both a product of nature and an artifact" (Warren 1997, 117). It was transformed by earlier cattle raising on it and later by the conservationists' activities to preserve it (Warren 1997, 116–117).

Very little of the earth's surface has not in some way been anthropogenically altered. As one geographer argues "there are no virgin tropical forests today, nor were there in 1492" (Denevan 1992, 375). Though the claim is controversial, with evidence for both sides, many archaeologists claim that not only did people occupy the Amazonian rainforest in larger, more densely populated clusters than was once believed possible, but their occupation may have been sustainable and they may have actually improved the environment of the Amazon basin (Mann 2000a, 2000b). Another argues that the European colonists to the New World wrongly believed that they had encountered the "forest primeval." Budiansky explains:

> One of the great ironies of the forest primeval is that the dense, thick woods that the later settlers did indeed encounter and arduously cleared were not remnants of the "forest primeval" at all. They were the recent, tangled second growth that sprung up on once-cleared Indian lands only after the Indians had been cleared or evicted and Europeans had suppressed burning (Budiansky 1995b, 106).

The "virgin forest" found by American colonists migrating westward "was not encountered in the sixteenth and seventeenth centuries, ... it was invented in the late eighteenth and early nineteenth centuries" (Pyne 1982, 46; see also Budiansky 1995, 106). Krech as aptly commented that the so-called "virgin lands" would more appropriately be called "widowed" lands (Krech 1999a, 99). Consequently, "paradoxical as it may seem, there was undoubtedly more 'forest primeval' in 1850 than in 1650" in North America (Rostlund, 1957, 409). By the 20th century, it should be obvious to everyone, but unfortunately it is not, that there are no "virgin forests" and wildlife left in the United States, though campaigns are still waged to preserve them. Over a half century ago the conservationist Aldo Leopold recognized this, though many who claim to follow him still deny it. "Every head of wild life still alive in this country is already artificialized, in that its existence is conditioned by economic forces. Sane management merely proposes that their impact not remain wholly fortuitous" (Leopold 1933, 21). Fire was a major tool used by the pre-Columbian American Indians to transform the forest so that it was more useful for them and also served to clear the underbrush, increasing visibility and therefore providing some protection against ambush by a rival group (Weatherford 1991, 41–46; see also Kloor 2000; Jacoby 2001).

> Early 19th-century travellers to America's western frontier sent back tales of spectacular scenery, frightening native populations, and bountiful mineral resources. Their reports fascinated the public, ignited the

entrepreneurial instincts of railroad builders and merchants, and enticed hordes of adventurers westward (Reif 2001).

Even that seemingly most pristine and untouched environment, the American West before European settlement, had in fact been profoundly changed by the American Indian as one would reasonably expect since they inhabited it.

Recently, scientists have challenged the ecologically pristine image of the American West. They claim that, by the time the early European settlers arrived, much of the landscape had already been considerably altered by native Americans over hundreds of years through burning, cultivation and settlement (Bond 2001).

All this conflicts with the image that has been so thoroughly ingrained in us in Hollywood movies and in my case, in what I learned every day growing up in the region. "The West was not as wild as everyone supposed, and Western films, apart from those set entirely in the desert, were played out against a backdrop as humanized as today's cultivated prairies" (Bond 2001). The myth of the "untamed West," as with other myths that we explore, would not be so pervasive and consistent, if it did not serve a purpose or in this case several different agendas. "Nature as it is represented in Westerns has not been affected by mankind. It is something other. It has a kind of purity and innocence. It is untouchable. That is what gives it its power" (Tompkins 1992 cited in Bond 2000). But it also gives power, purity, and heroism to the cowboys, the settlers and the cavalry that tamed it. The hardships of these pioneers— "windswept prairie, intense cold, plagues of insects, 250 days of frost"— often exaggerated but still very real, are almost beyond the imagination of those of us who have spent most, if not all, of our lives in the second half of the 20th century and now the opening years of the 21st century (Bond 2000). Yes, the hardships were very real, but if the land had previously been "tamed" then the pioneers were not conquering nature, rather they were conquering those who had already conquered, tamed, and transformed it, the American Indians.

These images of the pristine West also served the interests of those who wished to conquer and control the land. "Government and commercial expeditions were launched in keeping with America's belief that it had the right and obligation to bring civilization to the wilderness and its so-called primitive inhabitants" (Reif 2001). Perversely, the myth of the untamed West now serves the purposes of those who mythologize the American Indian as living lightly on the land. Both interpretations of the myth are de facto claims for the right stewardship of the land; one, because they (or their progenitors) tamed it, and the other by the environmentalists presuming to speak on behalf of the First Americans or using them to claim a type of stewardship that is in tune with "nature."

Our conception of the "wilderness," as Cronon argues, is an illusion or cultural invention. "Seen as the original garden, it is a place outside time, from which human beings had to be ejected before the fallen world of history could properly begin" (Cronon 1995a, 42). These can only be the beliefs of those who have the benefit of advanced technological civilization. However benign such romantic views of nature or wilderness may seem, they do in fact get in the way of intelligent problem-solving.

In its flight from history, in its siren song of escape, in its reproduction of the dangerous dualism that sets human beings somehow outside nature—in all these ways, wilderness poses a threat to responsible environmentalism at the end of the 20th century (Cronon 1995a, 43).

The stereotype of Indians as "pure lovers of nature" is only benign relative to the more monstrous stereotypes, but in other respects may be as destructive of their well being. "Natives" living "lightly" on the land was not grounds for praise but an excuse to take it from them, since they were not "using" it. What has happened is that the colonial mind has been reinvented by the environmental movement. It served colonial purposes to depict the conquered peoples as noble, leading a life of serenity and dignity, and doomed to die out. These characteristics, as with case of the North American Indian, became dominant only after they were defeated, though they had a long history. While still defending their land, there was no pejorative too vicious to attribute to them, summed up in the famed statement that the only good Indian is a dead Indian. Their dignity was in their dying, and it is precisely because they were allegedly dying with dignity that their conquerors are absolved of any blame for their passing. One author has sarcastically noted to me that Americans loved to watch the Indians die. The "vanishing native" is no longer a threat to our appropriation of their culture or to our economic and political dominance, so we can eulogize them and appropriate whatever material or non-material items of value are left. Commenting on the presumed acceptance of ethnic diversity, provided it does not interfere with economic progress, Dilworth observes that "as long as Native Americans and Hispanos could be perceived as powerless (or disappearing), they could be seen" (Dilworth 1996, 151).

The 19th century "Indian oratory" is being recycled by the environmental movement as a lesson for modern life. Unfortunately, some members of a conquered people who have been stripped of their land and independence will cling to a kind of spirituality that is falsely attributed to their ancestors and thus lends a superficial credence to romantic observers (For an egregiously bad romantic vision of pre-Columbian American Indians, see Sale 1990). These stereotypes of American Indians have evolved over the centuries of contact. Many are "holdovers from early Indian–white relationships" (Trimble 1988, 188).

[T]hey range from the benign observation that Indians are untamed, innocent, and pure lovers of nature to the more caustic description of Indians as savages, animals, and murderers. During more recent times, the tone of the stereotype has calmed down somewhat, but although the Indians are viewed in more passive terms, the stereotype remains (Trimble 1988, 188).

Deborah Root speaks of Western culture being "permeated with the duplicitous, Christian notion of victimization, which, on one hand, implies a moral or spiritual superiority and, on the other, a weakness that must be overcome through various spiritual struggles" (Root 1996, 100). She adds:

The white fascination with the romantic, abstract heroism of Native people is able to function as another means of colonial pacification because it presupposes the inevitable defeat and disappearance of the nations (Root 1996, 100).

Frederick W. Turner III describes the elements of the "new stereotypic image that we have made" of the Indian. One myth is that "the Indian was the original ecologist, killing only what he needed, caring for the natural world through which he moved" (Turner 1977, 10). To Stewart Udall, the Indians were the "First Americans, First Ecologists" (Udall 1972, see also Callicott 1991, 242; for a devastating critique of this thesis, see Krech 1999a). It is easier to live lightly on the land when you are few in number. Waste was unthinkable to many Indian groups "since tribal existence was often at a little more than a subsistence level and was occasionally less than that. Many tribes, such as those of the upper Midwest, suffered severe seasonal deprivation" (Turner 1977, 11).

There is a vast literature on the overkill hypothesis—namely the extent to which, throughout a long period of the habitation of North America, there was killing beyond need and human-induced extinction of numerous animal species (see Martin, 1973 for the article that originated much of the debate; Martin and Klein 1984; Klein 1992; Stuart 1991 for more recent comprehensive accounts and bibliographies). Martin's 1973 article raised the possibility that the autochthonous inhabitants of North America rather quickly hunted to extinction numerous species of large animals that inhabited the continent at the time of their arrival (for a contrary arguement, see Grayson 1977). Certainly, this was not the case for the Plains Indians; the arrival of the horse and the gun led to slaughter of various kinds, including stampeding bison over cliffs (Isenberg 1993, 2000). And where the population was dense, such as in central Mexico, environmental degradation did occur (Stevens 1993). A further debated thesis is that wildlife preservation in North America prior to the post-Columbian conquest was facilitated by "war zones" between tribal groups which became "game sinks" or sanctuaries for wildlife. These were areas in which few hunters would venture because of the possibility of being attacked by a rival tribe (Martin and Szuter 1999; see also West 1995; Stevens 1999).

Similarly, for Africa, one author who strongly condemns the European conquerors for their destruction of wildlife nevertheless cautions against "the polar opposite view about Africans and wildlife: that they lived in beautiful harmony and if the white man had never come, the peaceful coexistence would have continued into eternity" (Bonner 1993b, 44; see also Adams and McShane 1996, 239).

The Legend of Chief Seattle

Over the last 30 years, the words of Chief Seattle (also spelled Sealth) of the Dwamish have been circulating about as a lesson in ecology for us all. He has become the successor to the counter-culture for the mystical/mythical teachings of Carlos Castaneda's creation, the Yaqui Indian, Don Juan (Clifton 1990, 232–233). One newspaper claims that Seattle's statement "has been described as the most beautiful and profound statement on the environment ever made" (Houston Press 1990). Peter Sculthorpe, one of the most innovative and original composers of our time, was even creating a string quartet that deals with Chief Seattle's mythical "letter" (Uscher 1990, 51). Among those extolling the virtues of

Chief Seattle's "letter" are Paul and Anne Ehrlich (who seem prone to accept every romantic myth, from the Tasaday, to the parchuting cats in Borneo [see page 94], to the legend of Chief Seattle). The Ehrlichs' tribute to the mythical letter of Chief Seattle immediately follows a sentence in which they urge us to learn the "secret of people living together ... buried in the culture of the gentle Tasaday," a people whose existence as a culture is as fictional as Chief Seattle's "letter" (Ehrlich and Ehrlich 1981, 239; on the Tasaday see Berreman 1991, 1992; Yengoyan 1991; Headland 1992a, 1992b).

For Earth Day 1990, Chief Seattle's words sprouted up through the media. There were several difficulties with attributing those words to Chief Seattle. First, it was described as a letter, when in fact it was a speech. Second, Chief Seattle did not speak or write English. The original version of his speech was a copy and translation made by a Dr. Henry A. Smith who was present and related his account to a newspaper 33 years later (Bagley 1931, 255–256; Binns 1944, 100–104; see also Feest 1987, 1990; Vanderwerth 1971, 117–122; Armstrong 1971, 77–79). Consequently, we don't know exactly what Chief Seattle said or to what degree the account is accurate even in translation. The presumed speech does reflect the prevailing sentiment of the "Indian oratory" of the late 19th century.

There are just enough similarities in sentences or fragments of paragraphs to compare the speech in its original form with the current version. But the recent versions of Chief Seattle are also significantly different and reflect more the ideological needs of a contemporary movement, not the sentiments of a great mid-19th century leader. One of the most scholarly, perceptive, and complete studies of the "many speeches of Seathl" suggests that Dr. Henry A. Smith's presumed rendering of Chief Seattle's speech may have had more to do with issues in the 1880s politics of the city of Seattle, Washington than anything that may have been said by Chief Seattle over three decades earlier (Gifford 1998, 21–61).

Diligent research by Rudolf Kaiser has traced the twisted trail of the legend of Chief Seattle and its modern incarnation in the form of a letter. The original speech itself is highly questionable as to its authenticity. The "letter" was written by Ted Perry (then on the faculty of the University of Texas) as part of a script for a 1971 film on ecology. Perry's script did not identify the author as Chief Seattle, but merely as a generic 19th Century Indian chief. The letter authored by Ted Perry has become one of the most widely translated, reprinted, quoted, eulogized documents in our time. It has been called by a cleric "a fifth gospel, almost" and reprinted with portions of the Bible. There are films, shirts, songs, recordings, radio and television programs, and innumerable other forms in which the "letter" has been presented or praised. Until Kaiser's research, Ted Perry remained largely uncredited (Kaiser 1987, 505–526).

Chief Seattle's speech has so repeatedly been transformed to fit current ideological purposes that for some these later versions have greater authenticity than the original. In other words, Chief Seattle's "letter" is what we want Seattle and others to say about the environment in order to authenticate our own beliefs. To one historian the "original translation of Seattle's speech ... lacks some of the holistic language" of the modern renditions. "Regardless of what the chief actually said in 1854, it is significant that modern white Americans want to credit him with a biocentric philosophy" (Nash 1989, 247). One philosopher, J. Baird Callicott, was worried about a possible "backlash" from Kaiser's research and revelations. One would expect a philosopher to be concerned about the gullibility of those who uncritically accepted the validity of such an obviously flawed document (Callicott 1989, 35–36; for critical comment see Brunton 1992, 1995; Wilson 1992).

One wonders whether these historians, philosophers, or other academics would be as accepting of myths that are indifferent to the truth if the speeches of George Washington or Abraham Lincoln were so regularly distorted to promote a militarily aggressive foreign policy or the suppression of civil rights. Would they speak of the original lacking a global perspective or holistic analysis, or would they criticize the proponents of the distorted version, no matter how worthy or unworthy their motive may be? In this case, as in many other instances, groups dedicated to planetary salvation are given a license to distort history or manufacture data. In the case of Chief Seattle the hoax was accidental, for what was intended by Ted Perry was the speech of a generic Indian. But it happened to be what some people wanted to hear and believe.

In a more recent case, a hoax was deliberately created by a physicist who was able to publish scientific nonsense in a prestigious, postmodernist journal (Sokal 1996a for the article; Sokal 1996b for the hoax revealed). Sokal's article gave rise to an extended debate in which the postmodernists, secure in their ignorance of science, found him at fault, not their own naivete for publishing such nonsense. There was, and may still be, a Web site where one could access Sokal's articles and the responses of many participants in the controversy (Sokal and Bricmont 1998).

Since the 1950s there has been an industry of fabricating favorable quotes from Lenin to use against policies that one opposed. This has been condemned as should fabrications for any other cause. As we have noted previously, some of the more mystical world views and actual sayings of the renowned American Indian Black Elk, may be romanticized interpretations by the American poet who "recorded" them (Gill 1993, 34–35; Skeltenkamp 1993; see also Black Elk 1971, 1979). The implications of prevailing misinterpretations of Black Elk's thinking merit extended inquiry similar to that on Chief Seattle, even beyond the excellent book by Skeltenkamp.

Finding the true origins of Chief Seattle's "letter" took the diligence of a scholar like Kaiser, but falsifying the letter was easy. The question is why have so many been so easily misled and in turn misled so many others in perpetuating the myth? After all, an imperative of journalism and academic scholarship is to check and recheck your sources. On inspection there are obvious errors of fact. One simple statement in the "letter"—"I have seen a thousand rotten buffaloes on the prairie, left by the white man who shot them from a passing train"—contains a number of gross errors. First there is no evidence that Seattle could speak or write English or that he wrote a letter to the president. Second, in 1854 the first rail line had just reached the Mississippi. There were definitely none crossing the Great Plains until after the Civil War, by which the time Chief Seattle had died. Consequently, there was in 1854 no slaughter of buffalo from the trains on the prairie. (However, some historians now argue that the slaughter of the buffalo may have begun by Indians as early as 1840 in order to trade their hides for a variety of manufactured goods [Robbins 1999]). And finally, there is no evidence that Chief Seattle ever left the Pacific Northwest to see *anything* on the prairie.

The obvious errors in the text, and the relatively easy task of falsifying it, make the tale of Chief Seattle's "letter" more than a fable. It is instructive! To those committed to an idea or movement, accuracy and truth become secondary considerations.

Unfortunately, the Chief Seattle "letter" is one of innumerable errors of fact that are often repeated by those who presume to speak for the environment. Further, there has been a tendency in the media and academia to excuse the most gross errors of fact and their most egregious perpetrators because their cause is noble and just. It is understandable that those who romanticize other cultures are uncritical of their sources of information or are superficial in their understanding of other cultures or religions, since feeling, not thinking, is the motivating factor in their belief system. No matter how successful we are, individually and collectively, life is fraught with problems. To turn to another culture or "alternative" belief system for solving all of life's problems is simplicity itself; but in fact, it doesn't work.

What the environmentalists and various academics have done is to "commodify Indians and their heritage" (Clifton 1990, 16). The cultural heritage of Indians or other non-Western peoples is interpreted as a form of praise which serves the ideological purposes of the author or speaker. We all agree that wantonly killing bisons on the Great Plains by conquering people of European descent was an unmitigated evil. But when the Indians killed bison and "took only the most tasty and nutritious parts such as humps and tongues," leaving the rest was "not necessarily wasteful" but in fact opened or enlarged the "niche" for scavenging (Barsh 1990, 106). Or if the Aborigines in Australia regularly burned the land, then that is considered ecologically beneficial; but when the Europeans burned the land, it is no longer benign (Pyne 1991a, 1991b; on fire and the environment, Pyne 1997). (New evidence raises questions as to whether the fires set in pre-colonial Australia were as benign as Pyne suggests. These fires may have been responsible for the extinction of many Australian megafauna [Miller et al. 1999; Flannery 1999; Fox 1999a]).

It is interesting to note that when earlier people with contemporary, non-industrial descendants—American Indians, Aborigines, etc.— engage in an activity that is considered environmentally benign, they are identified by name. But when these same groups' actions may have been environmentally destructive they will be identified with generic terms such as "humans" or "human settlers" (Fox 1999a). It is as if we in the modern world share in the blame for all destructive behavior whenever and wherever it occurs simply by being human. But being human does not earn us any credit when things were presumably done right. In the United States the conservation practice of fire suppression on public lands to keep them pristine seems to imply a "perception that nature was without fire." This policy has led to environmental degradation rather than preservation (Warren 1997, 126).

Those in affluent societies who romanticize and find harmony in the less-than-utopian actual conditions of other peoples, have, as we have noted, "commodified" them. By this we mean that those pursuing an agenda in a developed country have appropriated and redefined the life-ways of another people for their own purposes. Some have argued that we have also "commodified" natural phenomena such as mountains and mountaineering (Johnston and Edwards 1994). Simply stated, many people are making money exploiting American Indian traditions, such as sweat lodge rituals, which they don't understand. "Natural healing and shamanism have become a billion-dollar business in North and South America" (Michel 1995):

> Tribal spirituality has caught on like wildfire in recent years with the spread of the New Age movement, which like the American Indians gives great respect to nature in its philosophy. Herbs and other natural items sacred to American Indians line shelves of health-food stores and are sold at festivals and street fairs (Michel 1995).

It should be noted that one of the "natural" plants in use in rituals is tobacco. For the New Agers with their phobias about "chemical" carcinogens, the thought must have eluded them that the ritual pipe smoking, if it is "authentic" and "natural," subjects them statistically to one of the most lethal carcinogens in modern society and the carcinogen that causes more deaths than any other.

The appropriation of beliefs and practices of other peoples, particularly those of American Indians, can be highly profitable for the entrepreneurs of the sacrosanct. A frequently used term for profiting from the artifacts, symbols, and beliefs of other peoples is the "commodification of difference" (Root 1996).

> Ancient Indian rites and traditions, like sun dances, vision quests and purification sweat lodges, have become staples of self-exploration used by New Age spiritual seekers, mostly in trendy affluent places like Marin Country, Calif.; Santa Fe, N.M.; Sedona, Arizona; and ... in Boulder [Colorado] (Johnston 1993).

Some American Indian groups have complained about those New Age groups who are profiting from Indian beliefs.

> A shopper in most any suburban mall today can find a store selling Indian symbols like dream catchers, rain sticks and kokopelli, the humpbacked Pueblo sign of fertility that has been transformed into refrigerator magnets and key chains (Johnston 1993).

"Many tribes, far from being flattered by the imitators, have denounced the movement as cultural robbery" (Johnston 1993). Others call it "cultural appropriation" (Root 1996, 70). The National Congress of American Indians has announced a "declaration of war" against these "non-Indian 'wannabes,' hucksters, cultists, commercial profiteers and self-styled New Age shamans" who are exploiting sacred rituals (Johnston 1993). Other poignant terms are "white shamanism" and "plastic medicine men" (Root 1996, 94).

> 'This is the final phase of genocide,' said John Lavelle, a Santee Sioux who is director of the center for Support and Protection of Indian Religious and Indigenous Traditions. 'First whites took our land and all that was physical. Now they're after what is intangible' (Johnston 1993).

Tina Talkington, an activist with AIM (American Indian Movement), argues that the very "spirituality" that the New Agers are now seeking was "outlawed" before the passage of the 1978 American Indian Religious Freedom Act (Willis 2000, 24).

> Now the very people who outlawed our religion want to steal it from us. ... We're picky because our spirituality is about all we have left (Willis 2000, 24).

Dilworth, who has studied the history of Anglo-Indian relations in the American Southwest, goes to the heart of the current situation. "New Age seekers continue to appropriate Native American spiritual beliefs and practices in an attempt to achieve spiritual and cultural authenticity" (Dilworth 1996, 3).

The "whites" may no longer be taking the Indian lands but they are buying or leasing them. "In some pueblos north of Santa Fe, where land can be leased privately, whites now outnumber Indian residents" (Johnston 1993). A similar pattern is emerging that is damaging Hispanic culture in Santa Fe, New Mexico, where "large numbers" of new migrants are forcing "profound and often unwanted change on distinct and long standing ways of life." As the affluent move into Santa Fe because they love the culture, they may succeed in destroying it. In Santa Fe County Hispanics were 65 percent of the population in 1970 and 45 percent in 1990. "Modest Hispanic" neighborhoods were transformed into "gallery districts," doubling home prices and skyrocketing property taxes (Gober 1993, 6). Because of the rise in property taxes:

> Some Hispanics who are native to Santa Fe no longer can afford to keep their family homes. "Imagine not being able to afford to live in the town of your birth, in some cases the very home where you were born" (Gober 1993, 6).

Even before the modern exploitation of American Indian cultures, there was the famous "Taos Circle" of artists who "specialized in paintings of the Old West, particularly 'traditional' Indian scenes, and in the process did much to shape how the mass public understood Indians" as being "separate from modern life." They were seen as "exotic, mystical, peaceful people, their lifeways seemingly unchanged from days gone by, without historical or social context." The artists were sponsored by the Santa Fe Railroad which used their work for tourist promotions (Warren 1997, 116; see also Dilworth 1996, 17–19, 60). Warren finds "similarity" in the "stereotyping" of Indians and of wilderness.

From the completion of the railroad through the Southwest of the United States to the West Coast, the American Indian was used as a tourist attraction. "The museum was the proper place for Indians; in the 'real' world according to government policy, they were supposed to be vanishing through assimilation. But Indian people were not assimilating or vanishing. They resisted and persisted" (Dilworth 1996, 51).

When American Indians decide to profit from using some of their land for a low level nuclear waste dump, conservationists suddenly lose their zeal for indigenous rights and local control (Johnson 1995). It has to be recognized that indigenous groups like American Indians differ among themselves on key issues, just like the rest of us. Many American Indians believe that they can profit from their land by allowing uranium mining, logging, or a nuclear-waste dump. Others within the tribe oppose such policies and often accuse the leaders who permit such ventures of "selling out" (Cray 1998). The important thing is that it should be the tribe that decides what is the correct action in terms of their economic needs and culture. Shepard Krech argues the "Indian people have had a mixed relationship with the environment" but are used by critics "for the sake of a narrative" to attack the "larger society as they absolve the Indians of all blame" for environmental sins. Doing so:

> They victimize Indians when they strip them of all agency in their lives
> except what fits the image of the Ecological Indian (Krech 1999a, 216).

"Frozen in this image," is the belief that "native people should take only what they need and use all that they take," and if they "participate in larger markets," it should be from "traditional" products, not from the oil and coal on their land (Krech 1999a, 216). Kevin Gover was nominated by President Clinton to be his "administration's chief official on Indian affairs" (Reeves 1997). "A Pawnee lawyer whose

Albuquerque firm represents Indians in environmental disputes," Gover claims that:

> Environmentalists quite often fall into the trap of paternalistic or romanticized approach to dealing with Indians.... They say 'The poor Indians don't understand the impact, so we have to protect them.' I have actually had environmentalists tell me 'Well, that's not the Indian thing to do' (Johnson 1995).

To view indigenous populations as living in some timeless utopia that has been disrupted by Western intrusion and has to be restored to its original pristine purity is an illusion that benefits neither the local people nor the environment. As the American environmentalist, Aldo Leopold, creator of the concept of a "land ethic" cogently stated, "wild things ... had little value until mechanization assured us a good breakfast and until science disclosed the drama of where they came from and how they live" (Leopold, 1966, xvii). Wise conservationists would put access to a good breakfast as an equal partner on their environmental agenda.

CHAPTER 6

Demystifying the Environment

We are beginning to overcome the mythology of what one author called the "pacified past" (Keeley 1996, vii, 3). Warfare, aggression, homicide, infanticide, and violence of all kinds can be found in human societies at all levels of development, from hunters and gathers to modern life (Wrangham and Peterson 1996). Violence is not unique to modern life, as many would claim. When homicide rates ranging from 19.5 to 28 percent can be found among groups in New Guinea and Australia it is reasonable to say that violence is endemic in these societies (77).

Not only is violence widespread in human societies, but also in the group behavior of the chimpanzee, one of our two or three closest relatives along with the gorilla and the bonobo in the superfamily Hominoidea. Jane Goodall, whose commitment to the chimpanzee is beyond question, observed chimpanzees engaged in almost every form of aggression from wanton murder of infants and children to warfare that involved the annihilation of other groups (Goodall 1986, 111, 503–522; Goodall 1990, 98–111; see also Teleki, 1973). Like mythologies of peaceful hunters and gatherers, there were similar beliefs about vegetarian chimpanzees so that early observations of "predatory apes" were originally described as being "unusual" and "atypical." Now it is recognized that hunting and meat eating, including cannibalism, are a regular part of chimpanzee behavior (Goodall 1986, 267–312).

Teleki argues that the myth of the vegetarian chimpanzee was "vital" to some theories of human evolution that needed to have a "plant-eating, tree-living, forest dweller as ancestor to the emergent primate novelty—meat-eating, ground-living, savannah-dwelling hominid."

Robert Ardrey, in some of the most eloquent and erroneous writing in pri-
matology, has argued that the forest ape was an "evolutionary failure,"
having become too dependent on the "forest both in his need for fruit and
in his need for boughs" (Ardrey 1963, 112). The vegetarian gorillas were
doomed to extinction and knew it. Their sexual and territorial instincts had
atrophied and at night they literally fouled their own nests (112–116).

In Ardrey's vision, the pathway led from vegetarian primates such as
chimpanzees and gorillas to the weapon-making, meat-eating killer
apes out of whom humankind arose. "The union of the enlarging brain
and carnivorous way produced man as a genetic possibility" (315).
Aggression and the "drive to maintain and defend a territory" were the
basis of man the social animal and the foundation of the "virtues of
human behaviour" (172–174). While "man is a predator whose natural
instinct is to kill with a weapon, ... the non-aggressive primate is rarely
called upon to die in defense of territory" (316). "The hunting primate
was free," including being free from "eternal munching." Meat eating
and the concomitant "capacity to digest high-calorie food meant a life
more diverse than one endless meal-time" (317). The door to a limit-
less horizon of human possibilities was opened.

As Teleki shows previously, the myth of the vegetarian chimpanzee
was a critical element in theories about aggressive, hunting, meat-
eaters, and human evolution. Teleki's position has been reinforced by
the emerging evidence that the proto-human *Australopithicus africanus*
were meat eaters before they were tool users and, though capable of
walking upright, they still were able to live in trees (Wilford 1999;
BBC 1999a; Fox 1999b; Vogel 1999; Spoonheimer and Lee-Thorp
1999). This could be interpreted to mean there was no *Homo sapiens*
transition to meat-eating because it had already taken place, and tool
using and a larger brain were not necessary for that transition to take
place. It has been argued by a number of anthropologists that the
energy needs of our larger brain required a diet of foods with highly
concentrated energy such as meat (Mithen 1996, 103).

The myth of the non-aggressive, vegetarian primate and the peaceful "primitive" has also been a staple of a romantic view of life that
Ardrey was attacking. This is yet another case where seemingly opposite or contradictory philosophies and perspectives use the same vali-
dating mythology as social charters and intellectual foundations for
their ideology and political advocacy. When presented evidence that
counters one's cherished beliefs, it is amazing the ways in which people can rationalize away unpleasant facts. Two authors, in confronting
evidence of meat-eating chimpanzees, decided that such behavior was

extremely rare, declaring chimpanzees were vegetarians. Not only are these meat-eating incidents "extremely rare," but these "apes were unusual and atypical of the species in general, living as they do in un-chimplike surroundings" (Morris and Morris 1966, 228). It is not clear what "un-chimplike surroundings" would be, since chimpanzees live there and have for some time. Until we have more comparative studies there is the possibility that early humans were less successful than their chimpanzee kin in deriving their basic nutrition from hunting (Lee and DeVore 1968, 4; Teleki 1973, 176).

The argument that chimpanzees are inherently vegetarians and became meat eaters only as result of human-induced environmental stress is simply an inelegant form of denial for those who cannot accept facts that undermine and conflict with their worldview. Absent humans, there has been and will continue to be climatic variations through time that can create considerable stress on various animals and plants in an environment. The question then remains, whether environmental stress that was not induced by humans could also lead to chimpanzees becoming meat-eaters? If so, how is that different from what we now observe? Is one "natural" and the other "artificial," and what is the meaning of this distinction? If environmental stress with other than human causes does not lead to meat eating by chimpanzees, why not, and what evidence can the deniers offer? The stress argument would appear to imply that they are eating meat only out of necessity. One would reasonably expect to observe some hesitancy or other sign of uneasiness or dislike rather than the gusto and relish with which they have been observed going about this task. If chimpanzees are inherently vegetarians, they have yet to be fully informed of that fact.

One need not be a believer in the thesis that violence is part of our nature—a thesis that this author does not accept—to recognize that there are at least some tendencies to violence that have to be understood and controlled in whatever type of society we chose to create and be a part of. In many respects, the romantic view of the vegetarian chimpanzees, particularly owing to a close evolutionary connection to them, is simply an extension into the primate world of the romanticized view of the human past.

Demystifying the Past

Raymond Williams, in taking an intellectual "escalator" to the past, gives us a sampling of English writers back to the Middle Ages. He found that in all periods there has been a sense of a lost Eden that had occurred in the recent past (Williams 1973, 9–12; see also Lowenthal 1985, 74–90). One might say that utopia was in the past and always has been. Williams then carried his inquiry back to ancient Greece and found similar sentiments of a sense of lost organic unity (Williams 1973, 13–34, 35–54). The alleged organic unity between the ancient Greeks and their environment existed in a landscape which was shaped by soil erosion and environmental degradation created by its human inhabitants (Runnels 1995, 96–99). Runnels criticizes the "widely espoused principle today that the destructive ecological practices of modern civilization are a new development" (96).

> The popular press frequently carries reports of people who advocate returning to the balanced and reverential regard they suppose our ancestors had for the natural world. The Garden of Eden is a primal myth of Western civilization, and it was preceded in classical antiquity by the belief in the Golden Age—a time, alas now lost, when humans lived in innocent harmony with their natural environment (Runnels 1995, 96).

Runnels cites P. B. Medawar in calling this type of thinking "Arcadian" (Medawar 1984). They are not "Utopian," because the writers of Utopian works—Thomas More and Francis Bacon—created places where science and technology were used for human betterment, in contrast to Arcadia in which "one of its principal virtues is to be pastoral, prescientific and pretechnological. In Arcadia, mankind lives in happiness, ignorance, and innocence, free from diseases and psychic disquiet that civilization brings with it." They are "living indeed in that state of inner spiritual tranquility which comes today only from having a substantial private income derived from trustee securities" (Medawar 1984).

The Lost Arcadia

One of the most imaginative romantic renderings of early peoples is the belief that there was in Europe a peaceful, matrilineal, egalitarian peoples who worshipped an earth goddess and who lived in harmony with themselves, with others, and with the environment (Gimbutas 1974, 1982, 1989). This Arcadian bliss was then disrupted by patriarchal militaristic Indo-European invaders (circa 4,000 to 3,000 B.C.). Needless to say, this view has appeal to some feminists who have put forward the thesis as a new gospel. It also has appeal to others, as these early communities were small and decentralized, and that was all their technology permitted. The thesis is not just about our past, but it purports to offer a guide to a peaceful "earth-centered" future in which we live in peace and equality. Unfortunately, though its original modern proponent, Dr. Marija Gimbutas, is widely respected for her technical work, most of her professional colleagues reject her thesis as not being proved (Steinfels 1990). A difficulty with the Gimbutas thesis is that second millennium B.C. gravesites that have been "recently excavated in Russia contain remains of women warriors ... female skeletons buried with weapons and bearing wounds inflicted by similar weapons (Ehrenreich 1999, 119). Another author adds that historically "cultures organized around war and displays of cruelty have had women's full cooperation" (Pollitt 1999, 123).

This Arcadian mind-set has been called "green fundamentalism" (North 1995a, 40). Emphasis is placed on "holistic" science, "personal experience and revelation" similar to the 17th and 18th century English protest movements that loathed the sterile scientism of the Enlightenment." The modern variant of this belief "releases believers from ordinary civic duty." This perspective of "angry fundamentalism" ignores "twenty years of quite good science-based progress in environmental policy making" (96, 97; see also North 1995b).

Despite this science-based progress, we persist in trying to promote environmental action by using a romanticized portrayal of ethnic minorities in anti-litter advertising. The United States advertisements showing a "generic Indian chief weeping" at the litter in parks has its "Philippine counterpart in a public service announcement featuring Aeta elders from Zambales in g-strings imparting environmental wisdom to Filipinos." Paredes adds that "the underside of being such a 'noble savage' is that one is still considered a savage" (Paredes 1997, 14).

Donald Lopez raises similar issues with respect to the romanticizing of Tibetan culture. The challenge that Lopez makes to the romanticizers is not over issues of human rights and self-determination which Lopez believes "all people of goodwill (when presented with the facts) would support, without invoking the romantic view of Tibet as Shangri-La." The romantic view may even convert people to the cause. But Lopez is "convinced that the continued idealization of Tibet—its history and religion—may ultimately harm the cause of Tibetan independence" (Lopez 1998, 11).

During the past three decades fantasies of Tibet garnered much support for the cause of Tibetan independence. But these fantasies are ultimately a threat to the realization of this goal. To the extent that we continue to believe that Tibet prior to 1950 was a utopia, the Tibet of 1998 will be no place.

We may be disillusioned to learn that Tibet is not the place we have dreamed of. Yet to allow Tibet to circulate in a system of fantastic opposites ... is to deny Tibetans their agency in the creation of a contested reality (Lopez 1998, 11; see also Lopez 1994).

With science and technology we have demystified our environment which has led us to mystify other peoples' environments. This process of mystification, as Bishop shows, lasts until we acquire more knowledge about them. "Travel writing," he tells us, "is not concerned only with the discovery of places but also their creation" (Bishop 1989, 3). However nice myths and fables may be, however much they may give meaning and coherence to our identity, they cannot serve as a basis for understanding and problem solving. We must never forget, however loosely the word ecology is used, that it fundamentally refers to, or should refer to, a scientific discipline, and not an ideology. The irony of Westerners traveling to distant places to discover the "other" is that vast numbers of people from these places have migrated to developed countries where they live and keep alive many facets of their culture. "Paradoxically, then, the Western elites spend thousands of dollars and travel thousands of miles to find what they already have" (Bruner 1996, 160).

Nostalgia for What Never Was

Currently many have a nostalgia for a close-knit, nurturing, nuclear family which once characterized the American way of life. Or at least we believe that there was a prevailing style of family life of this ideal type that now has been lost, but to which we must return. Stephanie Coontz examines the history of family life in the United States and, much like Raymond Williams, finds that there was no time in which the reality in any way compared to our golden age beliefs about it (Coontz 1992). Golden age mythology is more than just nostalgia; it is often used to promote economic and social policies that would not otherwise be supported. It can even be a smoke screen to obfuscate the real cause of a contemporary problem. Coontz's work finds that economics is a critical factor in contemporary family difficulties, a factor that those who promote the "family values" and the myth of a golden age are loathe to admit.

Mark Twain scathingly referred to the "enchantments" of the slave-owning American South (whose literary elite were influenced by Sir Walter Scott) that "sets the world in love with dreams and phantoms; with decayed and swinish forms of religion; with decayed and degraded systems of government; with the silliness and emptiness, sham grandeurs, sham gods, and sham chivalries of a brainless and worthless long-vanished society" (Twain, 1911, 328).

If the good life was always in the past, then of course it never existed. Even if there have been Golden Ages, we are still forced to address modern problems with modern means. New technologies and their consequences in terms of larger populations, and so on, have created new realities and problems that cannot be comprehended or overcome with the solutions of the past, however effective or ineffective they might have been in their time.

"Primitive Man" and Living in Harmony with Nature?

R. Terry Rambo examines the "conventional wisdom" that "primitive people live in harmony with nature whereas civilized societies wantonly degrade their environments" (Rambo 1985, 1). In *Primitive Polluters*, he investigates the impact on the environment of the Semang peoples of the Malaysian tropical rain forest. Because they are small in number he finds that their aggregate impact upon the environment is minimal, even though their practices such as burning trash or swidden agriculture are locally polluting. The reader of this fine monograph could equally conclude that Semang pollution, in relation to their numbers and level of living, is actually quite high. As we have noted elsewhere, modern technology may be polluting, but it is almost always less polluting per unit of output than less advanced technologies. Rambo wisely includes the air the Semang actually breathe as a critical part of the environment (48–51). From Belize to Borneo, one can find houses built on stilts of various kinds so that items (such as coconut shells) can be burned underneath, filling the residence with carcinogenic smoke that functions, among other things, in killing insects, often including malaria-bearing mosquitos.

If those living in environments with endemic malaria, such as Borneo, should try to use more benign ways of mosquito control, such as manufactured chemical pesticides like DDT, there are any number of horror stories about the dangers of their use that circulate like fabled urban legends. One incident in Borneo has a mosquito control program using DDT and causing the destruction of wasps, who controlled the caterpillars, which now, unchecked, devoured the thatched roofs of the people's longhouses (Jamieson 1985, 397–398). Over the years, as I have seen this story, I have written authors concerning their source and never received a reply. One account was by an author who claimed to have just returned from Borneo; which was coincidental since I had just spent the year in Indonesia and had visited every province on Kalimantan (the Indonesian name for Borneo). While there, and on subsequent visits to Malaysia (with two provinces, Sabah and Sarawak on Borneo), I have asked about the story and nobody had heard of it. Paul and Anne Ehrlich add to the legend. They have DDT becoming more concentrated as it works its way up the food chain, from house-flies who are eaten by geckos, who are eaten by cats, who die of DDT poison, giving rise to a plague of rats, the destruction of food supplies, and the spread of bubonic plague (Ehrlich and Ehrlich 1981, 78–79). According to the Ehrlichs, the "government of Borneo became so concerned that cats were parachuted into the area in an attempt to restore the balance" (79). One major problem is that there is not a political entity and a government called Borneo. The Ehrlichs' source does not refer to the "government of Borneo" but does relate the basic story, which they claim was told to them by a World Health Organization officer, who heard it from someone else (Harrison 1968). Elements of the story may be true, but the manner of presentation, the errors of fact, and the bizarre parts of it, such as parachuting cats into an area, all combine to cast doubt on it.

Even where fire and smoke for insect vector control are not deliberate, cooking (and heating in colder climates) and other activities tend to fill the dwellings of many people with smoke. This was as true for the ancestors of the contemporary affluent people of developed countries as it is for today's poor.

> Indoor particulate concentrations, probably the best single indicator of toxic (noncarcinogenic) effects, are twenty times higher in villages of developing countries than in households where two packs of cigarettes are smoked per day (Stansfield and Shepard 1993, 69–70).

Even though controlling all the possible variables for a study of indoor air quality in poor countries and its consequence for the health of its inhabitants is difficult, the evidence still strongly points towards indoor air pollutants being a major health hazard for the world's poorest citizens. Acute respiratory infections "are the leading cause of the global burden of disease and have been causally linked with exposure to pollutants from domestic biomass fuels in less-developed countries" (Ezzati and Kammen 2001).

> Several studies in developing countries have suggested that an increased incidence of pneumonia is associated with exposure to organic fuel emissions, although several studies have had problems with controlling variables such as socioeconomic crowding (Stansfield and Shepard 1993, 69–70).

These "traditional" indoor fuels have been advocated as "appropriate technologies" without regard for their deleterious impact on human health. Some of these traditional energy sources, such as burning wood, create a number of respiratory and eye irritation (including eventual blindness) health hazards (Smith 1983, 285; for an excellent survey of the problems of renewable energy, see Bradley 1997). "Besides producing smoke, wood, straw and dung fires give off nitrogen oxides, SO_2, carbon monoxide, and carcinogens. This can lead to acute respiratory infection and chronic bronchitis—conditions that kill some 4 million infants and children annually" (RICPQL 1996, 33). Another author gives an even larger list of polluting emissions from the combustion of raw biomass: "SPM, CO, NO_2, POM, aldehydes, benzene, phenol, cresol, toluene, etc." (Kamat 1998, 57).

Unfortunately, stoves designed to increase the efficiency of burning wood also increase the pollution. "The more efficient the stove, the more pollutants it releases. Indeed, in extreme cases the stove that is clamped down to make it barely stay alight ... may behave like a wood gassifier or pyroliser, heating the wood in the absence of air to produce great amounts of organic matter for discharge into the air." Among these pollutants are carbon monoxide, formaldehyde, and benzopyrene (Allaby and Lovelock 1980, 422). Some of these pollutants such as benzo-b-pyrene are not only irritants, they are also proven carcinogens.

Rambo found it "surprising" that "at the local scale," the Semang impact on the environment was "not invariably quantitatively less significant than ours, despite the immense differences in technological power between primitive and modern societies" (Rambo 1985, 78–79). He further argues:

> Although the Semang do not cause significant air pollution at the regional or global level, they achieve quite respectable pollution levels

in terms of the immediate life space of the individual and the household (Rambo 1985, 79).

Stated differently, from a strictly anthropocentric perspective, the first and most critical aspect of air pollution is the air that we humans actually breathe into our lungs. It may come as a complete shock to many to suggest the possibility that in general, the people in developed countries not only breathe in less polluted air than people in poor countries, but we may be breathing the least polluted air since not too long after humans acquired the ability to domesticate fire.

Burning of domestic fires and heavy smoking of cigarettes results in atmospheric contamination with noxious gases and particulate matter equalling or surpassing the norm in modern cities. If citizens of Malaysia's capital city of Kuala Lumpur were confronted with air pollution of the intensity normal in Semang households they would rise up in outrage over the terrible state of the city's environment, and they would blame it on modernization and capitalism (Rambo 1985, 79).

In addition to the very natural but toxic chemicals resulting from indoor fires, air can carry other natural particles that are harmful to human health. From the study of Egyptian mummies we find that many of them had suffered from "fibrosis of the lung" (sand pneumoconiosis) from sand particles, a condition that could well be present today with those who live in environments with frequent sandstorms (Sanderson and Tapp 1998, 53). They also "suffered from emphysema, or that they had lungs full of soot from oil lamps and fires." The same research found that "even pharaohs suffered from arthritis, and the combination of coarse stone-ground grain in the diet and blown sand meant that most people who reached middle age had teeth worn down to stumps and, frequently abscesses too" (Kennedy 1999). Historically the indoor air that humans have had to breathe was as bad as that previously described. In the densely packed houses "the occupants breathed and rebreathed each other's air, increasing the risk of respiratory diseases."

Animal exhalations, animal dander, and dried animal excreta were ... added to the indoor air, which could become heavy with methane and the decomposition products of urea. ... Living in close proximity with animals gave our Old World ancestors tuberculosis, influenza, measles, and smallpox among other diseases that followed Columbus into the New World (Garn 1994, 94).

Many have claimed that the New World was relatively free of disease, that "the common intestinal parasites were restricted to the Old World before Columbus, and that American Indians were free from their presence in pre-Colonial times" (Saunders et al. 1992, 117). It was argued that there was a "cold screen," namely a band of extremely cold climate across the northern parts of the New World that the first migrants had to cross. The "cold screen" prevented Old World diseases from crossing because it killed off those who had one of these diseases before they could carry the pathogens to the rest of the continent. "In effect, the traditional model assumes a culturally and demographically static native population, stabilized in a pure, disease-free natural environment" (117).

From Stanley Garn again we learn that:

Studies of skeletonal remains, latrine pits, defecation sites, etc. have shown that the Indians of North America were often parasitized well before 1492. Paleoparasitologists have demonstrated high levels of parasitism in American Indian coprolites from many parts of the United States.... So, we may conclude that no human group was without its load of parasites (Garn 1994, 93).

Other scholars have found "hookworm, whipworms and giant intestinal worms" in addition to body lice in pre-Columbian America (Reyman et al. 1998, 378). Also found in New World mummies and/or skeletons is evidence of conditions that are almost universally found in the remains of hunter/gatherers and pre-industrial peoples in the Old World or New. These include diseases such as osteomyelitis, which had virtually disappeared in developed countries by the end of the 20th century (so that most of us have never heard of it), dental pathology (linear enamel hypoplasia causing a thinning and cracking of the tooth enamel), and other conditions of the bones or teeth that reflect long periods of disease and/or nutritional deficiency that severely stressed the organism and disrupted the growth process. One study found that 70 percent of New World pre-Columbian mummies had suffered "bilateral pneumonia (bronchopneumonia or lobar pneumonia) as well as other respiratory diseases" (Allison 1984, 521).

A number of New World paleodemographers are critical of the thesis that the Americas were a "disease-free paradise" (Ubelaker and Verano 1992a, 1; Powell 1992, 41; Saunders et al. 1992). On the contrary, "populations had been transformed by high levels of infectious disease long before European contact and the historically recorded epidemics of the early 1600s" (Ubelaker and Verano 1992b, 280). European contact and conquest brought "new pathogens to the established infection load made possible by precontact patterns of social organization" (Saunders et al. 1992, 118).

Smallpox and other diseases brought to the New World are often spoken of as being "virgin soil" diseases in that they were new to the region and the population had not developed any immunity or experience in coping with them, a condition that greatly magnified the potential devastating impact. And the impact was devastating, though the magnitude of the impact is the subject of considerable debate (Meister 1976; Henige 1990; Henige 1998; Sanders 1992, 179). Those who consider the disease impact to be of catastrophic proportions are more likely to also believe in some variant of the "disease-free" thesis. For many areas on North America, particularly those relying on a diet largely of maize ("whose protein is of remarkably low quality"), the population impact of the newly introduced pathogens was undoubtedly worsened by the population being "nutritionally stressed" (Krech 1999a, 79).

Part of the belief of the New World being "disease-free" may result from what is termed a "distorted use of the epidemiological concept of 'virgin soil epidemic' by anthropologists and historians alike." It has been used to describe a situation where the population had never been exposed to a "particular pathogen, such as variola major" (smallpox). However, epidemiologists use it to describe populations in which the "organism has not been present for many years, if ever" (Saunders et al. 1992, 117). In other words, Old World had experienced "virgin soil" diseases as devastating epidemics had occurred, died out, and then returned again possibly several centuries later with the same impact, as if they had never previously struck. It is argued then that in a similar fashion, some of the pre-Columbian New World diseases struck and then returned, so that the European contact was not the first incidence of the New World "virgin soil" diseases as defined by epidemiologists. "By focusing on the novel, rather than the recurrent dimension of virgin soil epidemics, anthropologists have subtly and unconsciously reinforced the disease-free vision of precontact America" (117–118).

Saunders, Ramsden, and Herring argue that there is an ideological dimension to the thesis of a "disease-free" pre-Columbian New World.

The idea of the Americas and its inhabitants being pristine prior to contact has deep roots in the western European mentality and over the centuries has earned its keep by serving a number of political as well as psychosocial purposes....

It served to rationalize the course of White-Indian relations in the centuries following the European invasions of the Americas, and it continues to validate the paternal, dominating role of governments over surviving native populations (Saunders et al. 1992, 117).

Interestingly, the historic use of the disease-free thesis may have served the status quo needs of the conquerors, while it is currently also used by those critical of the conquest and the resultant civilizations. Not to accept the thesis of a populous, disease-free pre-Columbian New World, one risks "being branded anti Indian" if not racist (Krech 1999a, 84). More to the point, the disease-free thesis today is an integral supporting element in an antitechnology belief system that posits alternative utopian ways of life by returning to a more pristine past.

There was even "urban decay" in at least one city in pre-Columbian America well before the arrival of European conquerors.

> For centuries, Teotihuacan was one of ancient America's crown jewels, with magnificent buildings, powerful kings, and bustling markets. But by 600 AD it was in the grips of urban decay, with pollution and disease killing off young laborers in this ancient Mexican city (Holden 1999, 31).

In a conference paper by Rebecca Storey reported on in *Science*, it was determined that in a 400 year period in Teotihuacan after 600 A.D., "the number of skeletons belonging to teenagers and young adults increased by as much as 35 percent. Their teeth and bones revealed poor nutrition and infections" (Storey 1998).

> 'The culprits were pollution and poor sanitation,' Storey suspects. With no sewer system, citizens depended on seasonal rains to flush garbage away, but it would have piled up and rotted during the dry summer, causing stench and disease. The city ceased being 'a dynamic, attractive place' (Holden 1999, 31).

Of course, in order for there to be "urban decay," there must be urban centers. Not only was Teotihuacan an urban center, but in its time it was one of the greatest centers of civilization in the world. It can be argued that at its height, Teotihuacan was second to none in the world. This gets to the very heart of our inquiry. In finding malnutrition, disease, pollution, high infant mortality rates, and low life expectancies in the pre-Columbian New World, in the Pacific Islands, or in hunters and gatherers, the purpose is not in any way to deny the greatness of their achievements, nor to make any invidious comparison of these peoples and cultures to the heritage that produced this author. The central thesis of this book is that *all* cultures and civilizations were plagued by these conditions until the 20th century—our own and that of everyone else. The pre-Columbian New World was not spared. And the evidence would seem to indicate that most peoples are seeking the technological transformations that will allow them to participate more fully in a process that has in many areas overcome these debilitating conditions. Outside romantics wish to preserve other peoples as they supposedly were and still are. They therefore choose to ignore the toll in human lives that people in earlier times (and in the present) with prior technologies experienced. Though the intention may be worthy, the result, if successful, of the romantic misrepresentation is to preserve other lifeways for us to enjoy at the expense of the well-being of those condemned to persistent economic poverty. Though we must never confuse economic poverty with poverty of the spirit, there is simply no evidence for the widespread belief that cultural degradation is the necessary price of economic progress. On the contrary, we argue throughout this book that technological change continually offers new opportunities for peoples and cultures to advance themselves in all areas of human endeavor.

A standard criticism of "primitive savages" by their conquerors was that these "simple societies" were without history, unchanging, and timeless. They left the natural world unchanged, which was often the justification for taking their land and livelihood from them (Schrire 1984b, 7). Within this framework one could say that they maintained a condition of stasis and harmony with their environment. They were, in short, childlike, in need of protectors like us, and were not progressive. There was a hierarchy of civilization: We were at the top, and they were at the bottom.

It is an irony of more recent romantics that they have accepted this civilizational hierarchy of the colonial mentality and its alleged empirical foundations, and have turned it upside down, with highest honors going to those with the simpler ways who live "lightly" on the land and are in "harmony with nature." The life of the American Indians, Polynesians, or Bushmen is one of stasis and sustainability. Many of these perceptions come from the environmental and organic agricultural movements. As we noted in Konner and Shostak's observations, these contemporary romantics are projecting on other peoples the lifestyles and ideologies that they favor. Unlike the anthropologists from the 1950s and 1960s who romanticized the !Kung San (Bushman) lifeways, these contemporary romantics make no effort to study those they set forth as role models for modern society. In both instances it is a case of seeing in others what we feel is lacking in ourselves, be it uninhibited sexuality, nonviolence, secure family and social structure, or living in harmony with the environment.

For some, hunters and gatherers and other pre-industrial peoples have gone from being the original "affluent society" to being the first conservationists. Conservationist thinking and practices have been attributed to numerous peoples, either historical, such as North American Indians prior to conquest, or the contemporary Indians of the Amazonian rain forest. A number of anthropologists have tested these interpretations and found them wanting (McCay and Acheson 1987). "Applying Western notions of conservation and man/environment relationships to the interpretation of non-Western systems, past and present, can lead to serious errors. It may be ethnocentric to assume that where property rights are exclusive (to villages, clans, chiefs, or individuals), conservation is either the intent or the happy side effect" (14). To formulate an idea of conservation, there has to be the belief that human actions have an effect upon the environment and on the accessibility of resources within it. McCay and Acheson and the other authors in their volume did not find this consciousness. On the contrary, they found that a number of peoples believe slain animals have the ability to regenerate and return the next season (Brightman 1987, 136; or Krech 1999a for example). In fact after conquest some Indian groups resented conservation practices mandated by Europeans (Brightman 1987, 135).

The authors of *The Question of the Commons* apply several tests for conservation to the data. For example, it has been argued that when some large animals become scarce, people will hunt smaller animals to allow the larger ones to regenerate. However, if groups are hunting small animals and a large one crosses their path, will they kill it? If they do not and consistently do not, that is evidence for conservation. If they do kill it, then the evidence would indicate that the group is hunting smaller animals because they are more available. If so, the data fit an optimal foraging or efficiency model more than a conservation model (Hames 1987, 101; Stocks 1987, 111; for a number of analyses of early stone tool users, see Torrence 1989). The historical record is even more devastating to the case for conservation practices. Robert Brightman, in his article on the Algonquins, documents "overhunting and starvation," indiscriminate killing, allowing the meat to rot, and hunting to depletion (Brightman

1987, 121–125). The new hunting technologies brought by the Europeans did not change behavior but intensified it. The "fur trade provided incentives," new technology "increased pressure on animals," and the result was conditions that "inevitably led to game shortages" (126). One author argues that the Indians of the Northwest deliberately tried to exterminate some animals because they believed that they brought the diseases that were actually transmitted by the Europeans. This thesis is subject to some controversy, not over whether the animals were exterminated, but whether the disruption was a product of pre-Columbian belief systems or whether the extermination was simply the result of greater economic security and opportunity derived from new technology and increased trade (Martin 1979; Krech 1981; particularly Trigger 1981, 24–26).

CHAPTER 7

Technology and the "Primitive"

The camera has allowed anthropologists and film makers to create a vast number of first-rate ethnographic films of peoples who live in small groups. A sample of those from the !Kung Ju/'hoansi to the Yanomamo finds that the presumed "natural" morality of small groups is not immediately apparent. Conflicts of all kinds, some extremely brutal (such as those of the Yanomamo) have been captured on film. It is estimated that one-fourth of Yanomamo adults die violently. Other Yanomami blow "hallucinogenic stuff into each other's nostrils through a hollow tube to attain visions and contact hekura demons. This drug, which is also used by shamans when they perform curing ceremonies, may cause chromosome damage" (McElroy and Townsend 1989, 143, 175).

Among the Yanomamo Indians of the Amazon, the most prestigious items are simple products of modern industry that have become vital to their well being—shotguns, machetes, fishhooks, and eye medicine. "Like other Amazonian groups, the Yanomami have come to regard steel tools, aluminum pots, cloth and other manufactured goods as necessities" (Ferguson 1995, 62).

The reaction of the Yanomami to the "outsiders from a different place and time underwent a transformation" as white men who were once seen as the enemy "were culturally recast as good" and were welcomed in order to obtain "flashlights, knives and axes" (Kraut 1994, 12).

Loincloths and bare feet were abandoned for shorts and flip-flop shower
sandals. Sneakers and digital watches became highly desired objects that
Yanomami all but venerated. The encounter could not be undone. The
last transaction in the Columbian exchange was concluding (Kraut 1994,
12).

The larger ethnographic record is sufficiently mixed to raise questions about any thesis of a biologically based moral code and a linear
inexorable decline in moral behavioral conduct with the growth of
technology and society.

The Kayapo Indians of the Amazon are an example of a people who
have been empowered economically and who have used their empowerment in ways that have brought consternation to those who once
romanticized them. It may spoil our image of the Kayapo to learn that
they own an airplane and have a variety of modern technologies such
as video cameras, which have become an integral part of defending
their heritage as they understand it. These technologies are instruments
of empowerment. The Kayapo are no longer "dependent on the outside
world for control over representations of themselves and their actions
but possess to a full and equal extent the means of control over their
image" (Turner 1991, 309).

According to one observer, "the idea of the noble savage has really
confused the issue.... (Brazilian) Indians, like everyone else, are oppor-
tunistic. In the past, they used resources sustainably, but not always"
(quoted in Holloway 1993, 93). Of course, this is true almost by definition for any group that continues to exist. According to Turner the
Kayapo see the natural environment as a means of their continued existence or as "part of the total process of producing human beings and
social life" (Turner 1993a, 527).

> They have no mystical sense or reverence or respect for individual trees
> or animals and feel no hesitation about chopping them down or taking
> them as game whenever their interests demand (532).

Fundamentally, the Kayapo see the Amazonian environment in which
they live as a source of their livelihood and their continued existence as
a people and thus they have obviously the most vital stake in its survival
without a need for any New Age mystification.

> What concerns the Kayapo is nature in the aggregate, or more specifi-
> cally, the survival and reproduction of a sufficient slice of the natural
> environment to support their traditional way of life (532).

If the Kayapo Indians of Brazil wish to sell logging (mahogany) and mining rights to their land, then their former "defenders" (in this case Greenpeace) attempt to prevent them from doing so (Holloway 1993, 93). Initially, a court in Brasilia put a ban on selling mahogany and "confiscated" the wood which it planned "to auction." But the Kayapo fought back using their newly acquired revenue and have been able to continue logging (Moffett 1994). For the Kayapo Indians who had "been trading in mahogany for more than a decade," the timber sales "allowed them to join the consumer society." It also provided them the money necessary to travel to the capital to defend their rights (Posey 1990, 14). Interestingly, organizations such as Greenpeace are often less successful in protecting forests in the developed countries or regions where they were founded, so there is a serious ethical question about their right to impose their policies on others who must bear the cost (*Economist* 1997a).

Economic exploitation of the rainforest in ways deemed unsustainable by environmentalists is not limited to the Kayapo. Latex collectors in Brazilian rainforests were strongly supported by outside groups because of the terrible exploitation that they suffered and because of the ruthless murder of one of their leaders Chico Mendes. Many "collectors in Brazil's Chico Mendes Extractive reserve—explicitly established to foster trade in non-timber forest products—now invest their profits in cattle ranching and forest clearance." Outsiders fail to understand that trying to earn a living from sustainably harvesting a rainforest is "very hard work" and that as "soon as people can, they get out of these activities." Knowledge of consumer products available in the "outside world" has spread even to seemingly remote areas of Borneo (which are really not as remote as outsiders wish to believe) where the people want "televisions, engines for their boats and manufactured products" (Pye-Smith 2001).

The Kayapo sold the use rights of their land "to generate income, which they used to purchase satellite dishes, helicopters, and urban apartments among other things" (Paredes 1997, 206). In other words, the Kayapo were acting like "politically incorrect Indians" (Paredes 1997, 206). The Kayapo forcefully demanded and received a "significant percentage" of the proceeds of a gold mine in their territory, and one village "used the first income from the mine to purchase a light plane and hire a Brazilian pilot." The plane was then used to defend their territorial rights in various ways, including patrolling "their borders to spot intruders and would-be squatters" (Turner 1993a, 535). Buying and learning to use video cameras to create documentaries etc., the Kayapo have been able to take to the world community the case for their rights on a number of issues of vital concern to them.

And it is now clearly "impossible for the Indians to return to their old way of life" (Homewood 1995, 5). The invasion of their domain has brought new problems to the Amazonian Indians that will require coordinated action by these groups to defend their rights and work with outside groups to facilitate joint efforts such as community health programs (Turner and Yanomami 1991). Mystification of the Indians or their environment by outsiders does little to help their cause. Nor does it help to attribute to the Kayapo some romantic, mystical identification with the environment that is benign and sustainable, and in line with the latest in environmental thought in developed countries. The recent history of the Kayapo and other indigenous groups fighting to retain or regain their rights shows that most groups are aware of the environmentalist agenda and rhetoric, and are able to use it (or at least try to do so) to further their goals and aspirations (Brunton 1995; see also Huber 1997). Unfortunately they are "used" by those who claim to be acting unselfishly on their behalf. If we wish to help various indigenous groups gain the full measure of their rights and still achieve environmental goals, then the package must include the ability of the people to become full participants in the modern global economy and consumer society, or at least participate to the extent that they wish to do so.

Before contact with Brazilian society, the Kayapo did not have a concept of having a "culture" but simply believed that they were acting as human beings (Turner 1991, 293–296). This is true, incidentally, of innumerable peoples around the world whose "ethnic identity" was formed in response to outside pressures and was not previously recognized as a common cultural identity. Outside influences not only helped to shape the culture, but those within the culture began to see their culture as a resource, as the Kayapo did. The fact that anthropologists come to the Amazon with resources to study the Kayapo has fostered the "awareness that their traditional way of life and ideas were a phenomenon of great value and interest in the eyes of at least some sectors of the alien enveloping society." The Kayapo were then "guided by a new level of consciousness of their 'culture' as a focus of their political struggle" (301–302). Speaking of the political use of video cameras by the Kayapo, Turner observes that:

> The acquisition of this technology, of both hardware and operating skills, thus became an important part of the Kayapo struggle for self-empowerment in the situation of inter-ethnic contact. Control over the power and technology of representation ... became a symbolic benchmark of cultural parity (Turner 1991, 306).

There are places in the world where economically and technologically less-developed peoples are being brutalized, their cultures ravished, their territory and livelihood stolen, and all too often they are wantonly murdered. Their being victimized by more technologically advanced people is not an argument for preserving their technological backwardness, as many seem to believe. The argument for extending some measure of protection is to give technologically less-developed people the freedom to change (or not to change) at their own pace and advance technologically on their own terms, which they will inevitably do if the choice is theirs to make. Given the opportunity to change on their own terms, very few, if any, will exercise the option not to change.

The fact that romantic notions about peace-loving peoples fail to hold up to critical analysis does not in any way even remotely justify vicious policies and actions against these relatively defenseless people. This cannot be repeated too often. Some well-intentioned defenders of Brazilian Indians have charged that those ethnographers who have documented violent behavior in Indian groups, were unwittingly abetting the violence. One would hope that on more serious reflection, those who make these charges would not wish the defense of the human rights of the Indians to rest on whether or not they are violent.

The "Natural" and the "Primitive"

In being romanticized by New Agers, less-developed peoples often lose their individual identity and become the anonymous Indian or African or Asian. Sally Price quotes from the *New York Times* news service about a mine disaster in South Africa where 177 workers, most of them black, were killed. "A statement from the mine owners, General Mining Union Corp., identified the five dead whites by name, occupation and marital status, giving details of how many children they had. The blacks were identified only by tribe" (Price 1989, 56; Schuettler 1998). We know the name of Sir Edmund Hillary and others who have gone to Nepal to scale Mt. Everest, but "we cannot summon up the names of the Nepalis who accompanied them" with the exception of Hillary's companion, Tenzing Norgay, whose name was misspelled in Hillary's memoir (Iyer 1999, 22). A similar discrepancy exists in naming those who trade in African art. "African traders outside of Africa" are referred "to as 'runners,' while European or American traders who go to Africa to buy are called 'dealers'" (Steiner 1994; for pre-Columbian American art, see Coe 1993, 273).

If the "tribe" is a timeless, unchanging entity, then individual creativity would be a contradiction, since it would imply change. "In the
Western understanding of things, a work outside of the Great Traditions
must have been produced by an unnamed figure who represents his
community and whose craftsmanship respects the dictates of its age-
old traditions" (Price 1989, 56). Thus Price titles a chapter "anonymity
and timelessness" because the two are intricately interrelated in our
conception or misconception of "primitive art." If a piece carries a
name it is frequently that of a famous Western owner. If a work is
anonymous and timeless, who can claim rights to it except those who
have acquired it? Primitive art is not only anonymous, but it has come
to "signify the mystical magic ritual zone prior to history" (Errington
1993, 222). It is interesting to note that the much maligned tourist art
has facilitated the emergence of individualism and experimentation in
African art (Jules-Rosette 1984, 139–140, 155, 158, 204).

The "Natural" and the Purity of the Past?

What is ignored by those who venerate the "simpler" lifestyles is
that the growth of technology is a growth in ideas and knowledge. And
if we are going to be able to address our contemporary problems with
intelligence, it will be through knowledge and understanding, not
through a mystical reverence for our supposed "natural" condition. It is
interesting to note that in the United States the most dangerous occupation is the one presumably closest to nature, farming. Not only is it
dangerous to farmers, but also to their families (Ingersoll 1989).

Those committed to a belief in a prelapsarian state of benign grace
or who imagine that "a glorious pastoral world has been lost, through
machines," should read the poignant observations of one of America's
leading novelists and writers, Joyce Carol Oates. This admonition
applies also to anyone who "identifies himself as a child of the city,
perhaps a second- or third-generation child of the city." An individual
"who has lived close to nature, on a farm, for instance, knows that 'natural' man was never in nature; he had to fight nature, at the cost of his
own spontaneity and, indeed, his humanity" (Oates 1973, 38).

It is only through the conscious control of the "machine" (i.e., through
man's brain) that man can transcend the miserable struggle with nature,
whether in the form of sudden devastating hailstorms that annihilate an

entire crop or minute deadly bacteria in the bloodstream, or simply the commonplace (but potentially tragic) condition of poor eyesight (38).

Oates adds that it is "only through the machine that man can become more human, more spiritual" (38).

Understandably, only a handful of Americans have realized this obvious fact, since technology seems at present to be villainous. Had our earliest ancestors been gifted with a box of matches, their first actions would probably have been destructive—or self-destructive. But we know how beneficial fire has been to civilization (Oates 1973, 38).

May we add to Oates' claim that it was common among gatherers and hunters to set fire to the grass and bushes. For example, aborigines in Tasmania considered burning to be a way to "clean" the land (Allen 1997, 31; Blainey 1976, 67).

Nature: Its Unromantic Past and Present

Many aspects of what we revere and consider as "nature" have not always been similarly viewed by earlier peoples whom we deem to have been "closer to nature." Nature to them was any number of life-threatening events, from drought or plagues, to storms or wild animals. Nature, such as the forests and the sea, were also home to other threats, such as brigands in the forest or imagined maritime beasts in the sea. To one author, there is "equal evidence that tribal people who are subject to the whims of the natural world are bound to a respect in which there is a good deal of fear. Living in harmony with nature may be a fancy way of saying that people seldom had a chance to lift their noses far from the soil" (North 1995b, 243). The more we advance technologically and are able to some extent to control "nature" or at least reduce the hazards, then the more attractive "nature" becomes. Ironically, "alienation" from nature is a condition for greater love for, and involvement with, it.

Many of us read Petrarch's poetic paean in praise of Mt. Ventoux in our literature courses and were told that his was the first work to speak of a mountain as something to be climbed, with a view to be enjoyed, rather than an obstruction for travelers to overcome or a protective shield against hostile invaders. Prior to recent centuries, it was widely believed or "known" that the Alps were home to dragons. As recently as the middle of the 18th century, many who traveled through the passes in the Alps did so with a blindfold "lest they be overwhelmed by the awfulness of the scenery" (Fleming 2000, 6). "They saw goiters and cretinism among the region's inhabitants (the former still common in remote Himalayan areas) and talked of an alien race of subhumans. Until the mid-1700s, Mont Blanc was known as Mont Maudit—the Accursed Mountain" (Hightower 2001, 7).

The pre-18th century European view of mountains was shared by people in other continents and has been transformed to modern mystification through the influence of climbers from developed countries. "The typical attitude of the mountaineer once was 'if only they were flat, I could plough them' has changed through working and guiding Western Alpinists." The "aura of mysticism ... seeps into the vocabulary and thoughts of local mountain populations" because some of them work as guides, and "is reinforced when the elite local porters and climbers are taken to Western countries for climbing workshops and tours" (Allan 1988, 14). These are also the mountains where the Western mind locates the Shangri-La of its imagination (Lopez 1998; Klieger 1997). A story told to me by an anthropologist colleague has a South Asian villager comparing his family troubles to a mountain by saying that at a distance they look majestic but up close they are full of rocks and rubble. Mountains look very different to the outsider than to those who have to earn a living from them.

Similarly, the sea and the shore have been seen as anything but benign. In *The Lure of the Sea*, Alain Corbain presents a lurid tale of all the horrors that were ascribed to the sea and the shore throughout human history. Some of the fears were mythical and irrational, but many were based upon genuine dangers. Major storms came from the sea. Swamps and marshes that were often located along the coast were breeding grounds for insect disease vectors. Today we love the fresh air of the seashore or the cool night air. Even to the present day, it is "bad air" which is seen by many peoples as the cause of disease and death. Corbain traces the changing European attitude towards the sea and the shore and the rise of perceptions positing the harmony between the body and the sea, as part of the process of "inventing the beach" in the 18th century (Corbain 1994; Lencek and Bosker 1998, XXI). At least one scholarly reviewer of Corbain's book finds that he overlooks earlier literature that takes a more favorable view on the sea and the shore (Weber 1994, 15). The reviewer recognizes that the negative writings

Corbain uses are substantial and the current romantic views arose with the Industrial Revolution in Europe.

The building of roads, railroads, and later steamships allowed large numbers of people to visit the sea, or mountains or "exotic" lands. Many today in our era of jet airplanes are nostalgic for these earlier forms of tourism. Yet in their time, writers such as Washington Irving pined for "the good old times before steamboats and railroads had driven all poetry and romance out of travel" (Withey 1997).

Many 18th and 19th century lovers of natural landscapes admired them for their "picturesque" quality, which meant that they literally were like a painted picture. Many viewed nature literally through picture frames or a "Claude" glass (Andrews 1989; Bryant 1989). Currently, the technology of film and video often makes "nature" more interesting than the actual experience. "The viewer is not aware of the days and weeks that pass as the film maker waits to capture a single scene, because tedium is edited out of the show." Also edited out are the "lice, fleas, ticks, worms and other parasites" that "infest the noble predators and beautiful birds. Wildlife seems to be everywhere." In the actual preserves themselves, the few lions or other animals are often surrounded by carloads of picture-taking tourists. Otherwise, "nothing seems to be happening" (Lutts 1990, 150).

Nature, Technology, and Modernity

In Western culture, our modern sensibilities about the worth of "nature" and its preservation, about the rights of other living creatures, and even about our attitudes against cruelty and our sympathy for the rights of other human beings, have developed largely in urban areas in the past few centuries. "Alienation" from nature allowed us to be sensitized to the feelings and conditions of other humans and other creatures. Survival in urban areas did not depend upon killing those animals that are predators on our livestock or those creatures, large or small, that would consume our crop or carry disease to plants or animals. The culinary habits of 18th century England are challenging to modern sensibilities. From Diane Ackerman we learn that the "idea arose that torturing an animal made its meat healthier and better tasting." She adds that "they chopped up live fish, which they claimed made the flesh firmer, they tortured bulls before killing because, they said, the meat would otherwise be unhealthy; they tenderized pigs and calves by whipping them to death with knotted ropes; they hung poultry

upside down and slowly bled them to death; they skinned living animals."
There was a recipe "for preparing and eating a goose *while it is still alive*"
(Ackerman's italics) (Ackerman 1990, 147; see also Cartmill 1993, 104;
Iwatani 1998; Stern 1998). "Meat animals were tortured to death in vari-
ous ways to make their flesh more tender and savory" (Cartmill 1993, 104).
Animals were also tortured for other than culinary reasons.

"Alienation" and later affluence gave us the luxury of seeing animals
as pets or as creatures that have an independent right to exist. Many
who revered "nature" in the 18th and 19th centuries had less regard for
the rights of its creatures than did the urban consumers of books or art-
work. The eminent British historian Keith Thomas wrote in *Man and
the Natural World* that even those who presumed to love nature didn't
treat it as we would expect today. From Thomas we learn that "in the
eighteenth century the first impulse of many naturalists on seeing a rare
bird was to shoot it" (Thomas 1983).

Many of those whose names adorn numerous species of birds would
not today win the praise of the Audubon Society. "It is a fact of history
that most nineteenth century ornithologists were cold-blooded
anatomists, with no great skill in field observation or love for birds ... and
that it is mostly an unedifying aspect of our dominion over nature that is
immortalized in the names of birds" (Mabey 1988, 273; see also Mearns
and Mearns 1988). Not until well into the 19th century, did the technology of photography allow bird books to be illustrated without first killing
the birds. One of the important aspects or significance of Audubon's
work was that he could draw dead birds that looked very much alive and
interesting to observe. Photography and other technologies changed the
way that we look at birds, and the rest of the world (see Line 1998 for the
variety of technologies used by modern "birders"). Before photography,
the only way for an artist to draw a bird was to find one and kill it. Most
of the famous and great bird books before the advent of photography
were written and illustrated by those who went out and killed the finest
specimens so that they could draw them, a point made by Robert Welker
in *Birds and Men* (Welker 1955; see also Lutts 1990, 149). Live birds
move too quickly to be captured by the eye but not too quickly for the
camera. Only a very good artist was able to use a dead bird as a model
and create a drawing of a bird that looked alive.

The Natural is Not Property?

To say something is natural is an implicit denial of the human intellectual content of other peoples' endeavors. This is more than just an

academic point, particularly when there are increasing economic claims for intellectual property. This denial gives rise to a variety of asymmetries and has implications similar to those of the myth of the unique scientific and technological capability of Europeans and their descendants. The music of Elvis Presley is intellectual property that continues to pay a return long after his death, but the music of the black artists which he appropriated was not apparently defendable intellectual property. There was no recognition and no financial reward in his lifetime for Solomon Linda (and very little since he died) who wrote the "most famous melody to ever emerge from Africa" (Malan 2000). His music became the basis for a number of top money-making pop songs in the United States, including the world renowned "The Lion Sleeps Tonight," yet Linda died "penniless at the age of 53, more than 20 years after he and his band, the Original Evening Birds, recorded the song 'Mbube' for the first time" (AFP 2000). "Shortly after their recording, 'Mbube' was obtained by folk singer Pete Seeger in the United States" (AFP 2000). Seeger's "version of the song, called 'Wimoweh,' achieved chart success in 1951" (AFP 2001).

> Over the past 60 years around 160 recordings have been made of different versions of 'The Lion Sleeps Tonight.' It has been used in 13 movies, half-a-dozen television commercials and a hit play, *The Lion King* (AFP 2001).

In August, 2001, 62 years after the original recording of "The Lion Sleeps Tonight," versions of the song "will be recorded by top South African groups Ladysmith Black Mambazo and the Elite Swingsters and released as a compact disc" to benefit Solomon Linda's family (AFP 2001). In the Caribbean much the same story is told about the appropriation of the Trinidadian song "Rum and Coca Cola," which had immediate success similar to that of "The Lion Sleeps Tonight," and for which it took several decades of litigation to get recognition of the creator's property rights in the song. In agriculture the seed or plant products of biotechnology research facilities are intellectual property, while the germ plasm they used in their research is not, even though it may be the result of a long heritage of intelligent breeding and selecting of crops for a people's agriculture (Juma 1989). The ownership rights of the plant's genetic heritage is a complex intellectual and legal problem that is too often oversimplified. Agreements to protect farmers' rights by creating an "international fund with mandatory contributions" from those using the genetic heritage of poor countries are being considered. But, "this fund is not intended to directly compensate farmers or less developed nations for providing genetic resources." Instead, "it will be administered by a multinational organization to implement a broad range of conservation and development projects in regions of crop diversity." At least one private sector firm has agreed to give a share of the profits from using "indigenous knowledge to a general fund for conservation" (Brush 1993, 662). If the farmers don't control whatever funds are generated, it is far from certain that they will benefit from their expenditure. The fund does not exist yet, so for now the question is moot. In July 2001, there was an Agreement Reached on Protecting Plant Genetic Resources: International Undertaking on Plant Genetic Resources (FAO 2001). Much remains to be done to protect traditional agriculturalist breeders' rights but clearly there is movement in the right direction.

Companies and scientists in the United States have obtained patents to use extracts from plants such as turmeric (*Curcuma longa*, also called tumeric) and neem (*Azadirachta indica*) as medicine, which have long had medicinal use in India. At least one of these patents was legally challenged by the Indian government in what they call blatant "biopiracy" (Agarwal and Narain 1996; see also Stock 1999; Kerr et al. 1999; Knight 2000; Mashelkar 2001). Part of the challenge was successful, as a patent for the use of turmeric as a healing agent issued to a U.S. university was canceled by the U.S. Patents and Trademark Office in August 1997. However, the patent may be reinstated for specific uses, such as surgical procedures (Sampat 1998, 8).

The efforts to have the courts cancel a patent to employ an extract of neem for medicinal and agricultural uses have not yet been successful in the United States, but the patent has been revoked by the European Patent Office, which has thus far granted 11 other patents based on the use of the neem tree (IATP 1997b; Hoggan 2000; Kirby 2000; Jarayaman 2000; Hellerer and Jarayaman 2000). The Indian government's "re-examination request" on a U.S. patent for a type of basmati rice resulted in a successful challenge of three out of 20 claims. The company withdrew some of its patent claims and others were rejected by the patent office, leaving three successful patent claims (Mashelkar 2001). While many in India viewed this as a victory, the Western NGOs and their Indian subsidiaries saw it as a defeat (Mashelkar 2001; Tata 2001, 13; Madeley 2001).

One author has challenged the accepted version of the story of patenting neem, calling it "alarmist" and "nonsense." According to David Richer, the patent in question was for a process of extracting neem oil and not for the use of neem itself.

Patents are granted only for inventions that are new and obvious, and the use of the neem seeds in pest control fell into neither of these categories (Richer 2000, 206).

Richer adds that a firm could not "prevent farmers from continuing to use their traditional methods of pest control" nor could any patent "stop anyone from doing something which he was doing before the patent application was filed" (Richer 2000, 206). Since the firm seeking the patent on a method for processing neem into a pesticide would have to buy their neem in the open market (they do not currently grow their own) there is a legitimate concern that they would outbid the local farmers and thereby deprive them of their traditional pesticide and all-purpose healing agent (Kirby 2000). However real this concern maybe, it is not a legal basis for denying a patent for a process of utilization and must be addressed by other means. For those farmers who utilize seeds from their own neem trees there is no problem, but rather only for those who obtain their neem from others. In the long run more neem trees would be planted and harvested if the demand warranted it, but this obviously takes time and as J. M. Keynes said, "In the long run we are all dead." Not only would more trees be planted, which is generally considered a good in and of itself, but also further utilization should be welcomed by the current critics of the patent in that it would make widely available a pesticide from neem, which seems to fit two of the critics' most sacred categories, that of being "natural" and of being "traditional."

For India much of this controversy might be avoided in the future as it is creating a "digital database of its traditional knowledge" with the intention of having it "included in the patent classification system of the Geneva-based World Intellectual Property Organization" (Jarayaman 2000, 267). This would be useful not only in sorting out patenting issues early in the process, but also in making that very valuable body of indigenous knowledge available for use elsewhere with licensing arrangements that benefit both the users and the people of India. The more such digitalized information is available for indigenous knowledge for the use of all peoples around the world, the better we will be able to address this complex problem of patenting.

Bolivian farmers have been fighting efforts in the United States to patent one of their traditional crops, quinoa (ANAPQUI 1997). In Brazil, legislation has been passed to protect the intellectual property rights of the indigenous population (Bernardes 1997; IATP 1997a; LaFranchi 1997). There is a continuing controversy over the patenting in the U.S. of ayahuasca, a plant sacred to the Cofan peoples of the Amazon basin of Ecuador (Davies 1999; Woodard 1999). And there is a legal battle over a patent obtained by a U.S. firm for a variety of bean that has long been grown in Mexico and exported to the United States (Dalton 2000). It should be noted here that there are similar controversies involving patenting key elements of genome sequencing and in other areas such as computer software. In all these newly emerging areas of inquiry and discovery, we have yet to devise rules which protect traditional rights, encourage innovation, and do not allow an earlier innovation to block later innovations or to collect an unjust ransom. These are not always compatible objectives, and working out the fine details is still a work in progress and is not unique to issues of agriculture and "biopiracy" (see for example, DePalma 2000). Ironically, those NGOs in developed countries who vigorously and vociferously fought for international treaties protecting poor countries against "biopiracy," now find that their efforts have led to a regimen that has stifled the exchange of plant breeding materials to the detriment of all farmers rich and poor alike, and which is now forcing them to work to undo the damage they have done (Charles 2001a, 772–775). Fortunately an international agreement has been reached that largely restores—some countries are not fully participating—conditions more favorable to the sharing of breeding materials in a way that attempts to be fair and benefits growers and consumers as well as those from whom the breeding material is obtained (Charles 2001b).

The larger issue of intellectual property has become a major area of contention in the negotiations over WTO (World Trade Organization). For example, developing countries do not wish to be too restricted in their ability to produce low cost pharmaceuticals for public health needs while those countries where the producers are located are concerned that without patent protection further developments will be thwarted and all will be harmed including those most in need of new cures (Harmon 2001; see also Debroy 2001). Part of the problem is that patenting of intellectual property in software and biotechnology such as in pharmaceuticals or agriculture is such a new area that it is not at all clear what the proper balance is between protection of existing rights and the freedom to innovate. Many of the first software and biotechnology patents were granted too broadly and based on vague and hypothetical utilities before anyone really knew the long-term implications of the invention and the patent. Patenting requirements in these areas have now been tightened in most countries but the early patents remain in force—often as barriers to subsequent development—even for those claims that would not today be patentable. Intellectual property rights in software and biotechnology will likely be issues of concern and contention for some time, and serious thinking, rethinking, and negotiation of differences is of utmost importance if development in these vital areas is to proceed in the interest of all of humankind.

Currently, there is also a major debate on the ethics, legality, and research implications for agriculture of patenting life forms (IRRI 2000a, 2000b). This practice is allowed in the United States and Japan, and has been approved by the Parliament of the European Union (IATP 1997a). Patentable life forms are generally those resulting from research and manipulation in a modern laboratory. As of yet, agricultural and other life forms long in use ("traditional" or "natural") are not considered patentable. In other words, one is considered "intellectual," while the other is considered "natural," as if the outcome of long-term evolutionary change by farmers were a historical accident and not the result of practical operating intelligence.

In these and other instances, the distinction between the intellectual and the natural (or traditional) is the difference between a legal and/or ethical claim to income and no claim at all.

The widely recognized need for agricultural practices that are sustainable has led to the appropriation of the word "sustainability" as an identifying buzzword for a variety of back-to-nature schemes. Without question the issue of sustainability is legitimate and important and is a central consideration for anyone concerned with developmental issues, particularly in agriculture. However, to some enthusiasts, sustainability has become a synonym for stasis. It also means that we will no longer have to acquire new knowledge and values and otherwise respond to a changing world. Though many proponents of sustainability have a self-image of being radicals on the cutting edge of change, they are in fact reactionaries seeking not to return to nature but to the womb. Too often there is the implication that the rest of us favor unsustainable agriculture. We must clearly distinguish between the advocacy of sustainable agriculture as ideology and the serious scientific research in this important endeavor. There has been an unfortunate tendency for advocacy groups to appropriate terms such as "sustainable agriculture" or "Integrated Pest Management," claiming them as their unique intellectual property, giving them a very narrow ideological interpretation, thereby undermining the very important scientific investigation and implementation of these principles to address real problems.

In these and other instances, the distinction between the intellectual and the natural (or traditional) is the difference between a legal and/or ethical claim to income and no claim at all. The return to a "natural" talent, be it the ability of an Elvis Presley or of a great athlete, would be considered rent in economics. It should be noted that in the modern free market economics, as well as in its classical predecessors, "rent" is the only form of income one can legally acquire in a free market that is considered under some circumstances to be unearned. In addition to returns to the "natural" fertility of land, our economics textbooks refer to the returns to the natural ability of athletes, or to the income of a high-priced, tenured colleague who is no longer publishing, as "rents."

Just as some romantics exalt the "primitive" and seek to protect it against the degradation of civilization and change, there are also the traditionalists in our own culture who, like their many predecessors, deny the "ultimate" value of anything outside Western civilization. To be worthwhile it has to be ancient (a "classic") and ours. In literature, it has been noted, the reactionaries have a strong, nearly exclusive preference for works by white males who have been dead for some time. It is incredible that in our time, writers must defend the fact that non-Europeans wrote great literature and books prior to colonialism (Adhikari 1988). Bruno Nettl, an ethnomusicologist, demonstrates conclusively that Western music and musical institutions have the same ritual functions as those of so-called "primitive societies" (Nettl 1992, 8–34). Just as the romantics would protect others from corruption by us, the reactionaries would protect us from corruption by others. One would have thought that such ethnocentrism died with the passing of colonialism and the understanding gained from ethnology and anthropology. Certainly, the experience of the 1930s and early 1940s provided lessons at a terrible price. Unfortunately in the last few years, a new ethnocentrism has emerged with a virulence. It is sometimes called "cultural fundamentalism" and is overflowing with ethnocentrism in the guise of enlightenment.

A deep and abiding commitment to one's own culture is not an impediment to appreciating another culture, but a precondition for it. To quote Franklin Delano Roosevelt's first inaugural address stating his "good neighbor" policy, the "good neighbor" is one who "resolutely respects himself and because he does so, respects the rights of others." Those who understand and appreciate their own cultures are best prepared to teach others about them and to learn from others about their cultures.

The worst reason to love one's own culture is because one has a mistaken belief that other cultures are inferior. Similarly, the worst reason to become enamored of another culture is because one is alienated from one's own. Those who find their own culture lacking will think that they find in another culture what they are seeking because it is missing in their own lives. Unfortunately, what they find is generally a figment of their own imaginations and not a genuine attribute of the other culture. And as we have shown throughout this book, outsiders with their own notion of what is traditional in another culture end up trying to impose this conception on the peoples themselves.

As noted elsewhere in this book, we have the New Age Romantics who bowdlerize American Indian culture, the pseudo-Buddhists who are "into Zen," and the back-to-nature enthusiasts and others who believe that hunters and gatherers and other poorer peoples lived, or still live, in "harmony with their environment" and had found authentic affluence. These belief systems, however well-intentioned, do not honor other cultures but demean them by portraying them as something other than what they are. In addition, though these New Age and postmodernist affectations may be absurd to the point of being humorous, they can, as we attempt to show, also be harmful and may be used as a basis for hindering the economic advancement of the very peoples they purport to honor and defend. And as we have argued throughout this book, we first project onto others what we find lacking in our lives or believe is lacking. When others fail to conform to our image of what their lifeways should be, too often we have opted to force upon them traditions that they have either rejected or simply never had.

What is fascinating lovers of the "natural" in the United States is how much reading, discussing, and attending training classes intellectuals must engage in to learn to do that which is natural. This includes giving birth to babies and for some also includes how and where to excrete bodily waste (See Ross 1988; Meyer 1989; Poore 1989, 58).

In criticizing many modern eating practices, such as too high a fat content in our diet, many an adverse comparison is made to the earlier, higher fiber content diets of our ancestors. The implication is that these traditional "natural" diets were, overall, better than ours. True, they were different from ours, but not necessarily better. Similar claims of superiority are made for the diets of poorer peoples today, particularly those people that we frequently define as primitive. Consequently, when people find means of earning cash income and thereby acquiring foods from outside sources, it is seen as a worsening of the diet as well as part of a larger degradation of a culture.

The Healthy "Primitive?"

There is a large body of anthropological literature arguing that prior to acculturation these people were "healthy and well nourished" in an "equation of success with a homeostatic relationship with the environment" (Denett and Connell1988, 274). This literature is now being challenged. For example, "reduced body size" was seen as an "adaptive response" to lower food availability; now it is also seen as "symptomatic of nutritional stress" (275).

In a study of peoples of the Highlands of Papua New Guinea, Dennett and Connell found many cultural practices contrary to the health and welfare of the population. It is true that when we study a people closely, many practices that may seem strange to an outsider turn out to be highly adaptive, but it is also true that many other practices are destructive of nutrition and health. Dennett and Connell found in the central Highlands of Papua New Guinea that "there is evidence that not only are nutrition levels improving over time with increasing access to money incomes, but nutritional status varies with degree of acculturation" (279). When we find that groups that are presumably living in ecological balance with the environment have high infant mortality rates with low life expectancies, then one has to question the presumed cultural virtue of homeostasis.

Before sustained contact with Europeans, among the Inuit (Eskimo) contagious diseases were rare because of the small size of the groups. However, that does not mean that they were in pristine good health. Instead "the health problems of Inuit were primarily chronic conditions such as arthritis in the elbows, eye damage, spinal defects and inflammation, deficiency in enamel formation on the teeth, loss of incisors, and osteoporosis" (McElroy and Townsend 1989, 31).

> Hunting hazards included snow blindness and sensory overload due to glare and isolation in a one-man boat, the kayak. There was a risk of contracting tapeworm and trichinosis. Eating aged meat, considered a delicacy, posed a risk of fatal botulism (31).

Living close to nature, as most of us would undoubtedly define the traditional Inuit way of life, does not mean that any significant portion of the population died of what we euphemistically call a death from "natural causes." In any case, dying of "natural causes" is probably not all that common in the animal kingdom, particularly among those defined as prey and is a luxury achieved largely by humans in developed countries.

By far the major cause of natural death was accidents, especially drowning or freezing to death after capsizing, but including house fires and attacks by sled dogs. Hunting accidents among men accounted for 15 percent of the deaths of a southern Baffin Island group (31).

Other causes of death deemed as "an important regulator of population" were feuds, murders, suicide, and infanticide (31). One can still have great respect for the San, Inuit, and other peoples for their cultures and their adaptation to harsh living conditions without mythologizing their lives as being "savage innocents."

CHAPTER 8

The Human Endeavor as a Creative Force

The Nature of the Natural

In the marketplace of the affluent, the word "natural" sells products and sells them at a premium price. Natural is quality. The word artificial is a pejorative. Of course, there is nothing "natural" about great works of art from Beethoven, Tagore, or Basho to Shakespeare, Orozco, Achebe, Coltrane, or Marley. Natural foods are presumed to be healthy and wholesome. How things become defined as natural and therefore worthy is not clear. In a society currently obsessed with too much fat in the diet, a cereal called granola is "natural" though it has 11.1 grams of fat per 3 ounce serving (without milk) compared to 3.4 and 6.2 grams for the same quantity of ice cream and beef, respectively (Hippocrates 1989, 12).

Many who try to "eat closer to nature" have devloped a preference for that which is raw or uncooked. Not only is "natural" and raw not necessarily better, but food processing is an essential component of food safety. In recent years, the failure to pasteurize apple juice has caused illness and death (DeGregori 2001). Many of the important transitions that humans made in food production, particularly for grains but also for manioc, depended on food processing for their full realization. Food processing has been historically a vital part of human development. "It was only after man had learned to use fire for cooking, about 40,000 years ago, that it became possible for him to take advantage of a greatly expanded food supply in the form of cooked vegetable foods" (Kakade and Liener 1973; see also Bittman 1994). "Some cultivars are quite toxic, unless properly prepared" (Garn 1994, 90).

Human use of food processing has been a necessary component for the use of plants and animals that became the basis for feeding the density of human population necessary for establishing civilization. Absent the use of fire for cooking, we would be limited to plant products like fruits and nuts, as the other vegetable matter of the plant would be indigestible. Vegetable matter is primarily composed of cellulose and raw starch, which we are unable to digest and from which we are unable to extract much nutrient. Milling as well as cooking is vital for the digestibility of grains. It takes heat to break down the cellulose structure of plant cells and to bring about the chemical change in starch to make it digestible (Bates 1967, 39). Several million years ago, a change in early hominid dentation made it difficult for them to break down "tough, pliant foods such as seed coats and the veins and stems of leaves" (Teaford and Ungar 2000). Marston Bates aptly argues that "cooking, then, can be looked at as a sort of external, partial predigestion" (Bates 1967, 39).

Modern Food Supply and Safety

Food contamination has been a fact of human existence throughout history. As a carrier of botulism or ergot and aflatoxins from the fungus *Aspergillus flavus*, the consumption of food necessary to sustain life has caused mass illness, blindness, and frequently, large-scale death (Matossian 1989). Even food uncontaminated by micro-organisms contains substances that would be considered a threat to human life were they used as food additives. The production of toxins by plants was an evolutionary adaptation in order to avoid being eaten.

It is ironic that the term "chemicals" is exclusively a designation for manufactured chemicals and is used to condemn food additives. Plants are also chemical factories and generate toxins in far greater abundance than the small quantities of manufactured chemicals applied to them for pest control. Many of the "naturally" produced chemicals are highly toxic and have very active properties. Some of these chemicals are for medicinal use and some as poisons. And some of these same "natural" chemicals have been used for both purposes depending upon the mode of usage, particularly the dosage. Critics of modern chemophobia often quote the medical adage that dose makes the poison.

There are trade-offs and choices to be made in all areas of the human endeavor, with possibilities for substantial gain, but with no alternative totally free of some risk or another, and it is the same with advances in science and technology. Undoubtedly the yield, whether it be per land or per labor unit, has been a primary consideration for food production; nevertheless, other considerations in selectivity make domesticated plant evolution a complement to processing in making food accessible to humans. Such foods can in no meaningful way be called "natural."

Raw and Pure?

Though touted, sold, and eaten as "health food," raw alfalfa and clover sprouts were found in one study to be responsible for over half the food-borne illnesses causing outbreaks of *Escherichia coli* O157 and *Salmonella* (Mohle-Boetani et al. 2001; Huget 2001). "People with *Salmonella* or *E. coli* O157 food poisoning had consumed alfalfa or clover sprouts about 5 to 10 times more frequently than people without food poisoning." "As currently produced, sprouts can be a hazardous food. Seeds can be contaminated before sprouting, and no method can eliminate all pathogens from seeds" (Mohle-Boetani et al. 2001). The problem "is worse for sprouts than for other plants, though, because the seeds are incubated and sprouted in just the kind of moist, humid environment that *E. coli* O157 and *Salmonella* bacteria love ... You can't just wash the problem away" (Huget 2001). Irradiation of sprouts is the most effective way of ridding them of micro-organisms but tragically it is opposed by those who purport to be promoting food safety. The author's advice is clear:

The general public should recognize the risks of eating sprouts, and populations at high risk for complications from salmonellosis or *E. coli* O157 infection should avoid sprout consumption (Mohle-Boetani et al. 2001).

Contrary to the critics' claims, irradiation of foods such as alfalfa sprouts to protect against microorganisms may also preserve the quality of the food compared to the control. "Antioxidant power increased linearly with radiation dose at both 1, 7 and 14 days of storage" as irradiation had a "minimal effect on" total ascorbic acid content. In addition, "carotenoid content" of irradiated sprouts "was higher than the control at 7 days of storage" (Fan and Thayer 2001).

The arguments against food irradiation are so similar to those used against pasteurization that one contributor posted a piece on an agricultural news group in opposition to food irradiation. At the end, he apologized for a bit of trickery and explained that it was actually from a 1920s article and all we had to do to replicate it exactly was to replace the word "irradiation" with "pasteurization" and "food" with "milk."

Until very recent times, milk has been anything but pure. "Before pasteurization and refrigeration, brucellosis, undulant fever, and bovine tuberculosis often came with milk, a danger for children especially" (Garn 1994, 90). Pasteurization of milk took a half century to be widely accepted in the United States because of antiscience opposition similar to the current opposition to food irradiation by groups claiming to be defending consumer safety when they are in fact opposing what is currently the most effective means of protecting the consumer from meat infected with *E. coli* O157:H7 (Tauxe 2001). From another article we learn that after "Pasteur invented the process for pasteurization of milk, he promoted the use of this process on wine, not milk? It was a businessman, Nathan Straus, who championed the cause of pasteurizing milk to reduce infant mortality. Against the opposition of doctors and the milk industry, Straus installed a milk pasteurization unit in a children's orphanage and measured the impact." The results would not surprise us today (Agbiotech Bulletin 2001).

> Infant deaths were reduced from 44 percent to 16! Opposition to pasteurization included concerns that the process would: conceal evidence of dirt, mask low quality milk, remove the incentive to provide clean milk, increase the price of milk, and take the "life" out of milk (Agbiotech Bulletin 2001).

The major problem with food irradiation is the term. We have had over a half century of fears about anything involving radiation unless used for medical purposes or if it is what is called background radiation—that which exists in nature—no matter how high it may be. Attempts to calm this irrational fear of food irradiation by calling it "cold pasteurization" have been met by vigorous opposition from environmental groups. In the public's mind radioactivity is associated with grotesque mutations and genetic damage of all kinds as portrayed in any number of Hollywood films. One wonders what those who oppose food irradiation would think of deliberately bombarding plants or their seeds with radiation with the intention of creating mutations? Actually this process began back in the 1920s when radiation was viewed as benign, if not beneficial, prior to the dreaded radioactivity of the atomic bomb. The environmental activists' leadership is probably aware of the use of radiation in plant breeding as the resulting varieties are as widely if not more widely used in "organic agriculture" where there is a need for greater disease- and insect-resistant varieties than in conventional agriculture (Leaver and Trewavas 2001, 744–745). The activist leadership wisely keeps quiet on the subject, as the facts would undoubtedly upset their troops in the street.

In terms of all the arguments used against food irradiation or genetically modified food, mutation breeding should be even more vigorously opposed except for the fact that it has been done for so long, has been so beneficial, and without the slightest evidence of any harm whatsoever.

The application of gamma rays and other physical and chemical mutagens for crop improvement in the past 70 years has increased crop biodiversity and productivity in different parts of the world. The number of officially released crop mutant varieties has already exceeded 2200 (FAO/IAEA 2001).

In June, 2002, there will be a special "Symposium on the Use of Mutated Genes in Crop Improvement and Functional Genomics" in Vienna, Austria sponsored by the Food and Agriculture Organization of the United Nations (FAO) and the International Atomic Energy Agency (IAEA). The Symposium will attempt to "inventory the use and economic impact"—can you believe it?—of "super mutations" for the "improvement of crop production and address this message to plant breeders" (FAO/IAEA 2001). Fortunately, the meeting is unlikely to draw huge crowds of protestors, though in terms of the presumed logic of previous demonstrations, it is not at all clear why not. In Italy, the militant Green-antigenetic-modification-of-food Minister of Agriculture was apparently embarrassed when it was learned that one of the most popular varieties of durum wheat being used to make pasta was the product of mutation breeding (*Nature* 2001).

When the UNDP (United Nations Development Programme) published its *Human Development Report 2001: Making New Technologies Work For Human Development*, recognizing, in a very balanced report, the potential benefits for poor countries of genetically modified crops, the critics were unfazed and responded by vilifying this fine report (UNDP 2001). Since many of the mutated crops are already being grown in developing countries, mutation breeding cannot be so vilified, nor can the argument be made that it has done nothing for the poor.

A large number of these varieties are food crops released in developing countries. Some of them were obtained as infrequent mutation of specific genes responsible for agronomically important plant characters. This has resulted in the widespread use of these mutated genes in plant breeding programs throughout the world and has brought about an enormous economic impact, e.g. in barley, sunflower, soybean, rice and many other crops (FAO/IAEA 2001).

"Chemicals" and "All Natural" Cancer

There is a steady drumbeat in the media about alleged cancer-causing chemical additives. The often unstated presumption is that "natural foods" are free of "chemicals," although what exactly that may mean is unclear, since plants do in fact consist of chemical constituents. Breast cancer heads the list of the cancers that are alleged to be caused by "chemicals" with DDT, DDE (1,1-dichloro-2,2-bis[p-chlorophenyl] ethylene, the metabolite of DDT) and other organochlorines considered the worst culprits. Previously, I cited an article that listed eight studies that failed to confirm any link between breast cancer and organochlorines to which I added a ninth, which was published after the article that I cited (DeGregori 2001, 138). To these we can add a tenth study, this one taking five previous studies and presenting a "combined analysis of these results to increase precision and to maximize statistical power to detect effect modification by other breast cancer risk factors." Their conclusion was once again that the "combined evidence does not support an association of breast cancer risk with plasma/serum concentrations of PCBs or DDE. Exposure to these compounds, as measured in adult women, is unlikely to explain the high rates of breast cancer experienced in the northeastern United States" (Laden et al. 2001).

The latest breast cancer study was presumably necessary because some said the previous studies "might simply have been too small and that their combined data might reveal such associations, at least for some subgroups of women.... [T]hat explanation was dashed as scientists analyzing the combined data also concluded that neither exposure explains the high rates of breast cancer in the U.S. Northeast" (NIEHS 2001).

The term "chemicals" has become a code word for all that is wrong with modern life. Bruce Ames has long argued that we ingest far more of the major carcinogens from the foods we eat than from additives or other "chemicals" in our food or environment: "Despite numerous suggestions to the contrary, there is no convincing evidence of any generalized increase in U.S (or U.K.) cancer rates other than what could plausibly be ascribed to the delayed effects of previous increases in tobacco usage." Further, "there are large numbers of mutagens and carcinogens in every meal, all perfectly natural and traditional. *Nature is not benign.* It should be emphasized that no human diet can be entirely free of mutagens and carcinogens" (Ames 1983, 1261; see also Fumento 1999, 149). Ames has demonstrated that some of the foods we eat also seem to provide protection against cancer and help the body's mechanisms for neutralizing some likely carcinogens. Though Ames is seen as the enemy by "natural foods" enthusiasts, they have picked up his ideas about foods, such as broccoli, being anti-oxidants and therefore anticarcinogenic, and act as if it was their own discovery, or as if it was a validation of their ideas about human health.

Many non-scientists firmly believe that there is an underlying political agenda to the issues that we have been discussing throughout this book on chemicals and human health. This was the thesis that was clearly stated in the subtitle and throughout the text of a major book on the subject (Proctor 1995). If you believe that "chemicals" are dangerous, then you are moderate to liberal to left, while those who believe, following Bruce Ames, that "chemicals" properly used have brought enormous benefits to humankind are rightwingers beholden to the chemical industry, with very few moderates or even intelligent conservatives supporting their cause. It would come as a shock to most to learn that cancer researchers, when surveyed, described themselves as moderate (28 percent) to liberal (48 percent), with few conservative (17 percent). ("Leftist" is how Alan Sokal, previously noted as a scientist critic of postmodernism, describes himself. "Left liberal" and "a member of Democratic Socialists of America" are the affiliations of the editors of the anti-postmodernist book from which this survey is taken (Sokal 1998, 22).) Yet their view of the causes of cancer differ markedly from non-scientists of similar political views. In the survey which listed a number of scientists and asked whether they had high, medium, or low (or don't know) confidence in individual expertise on environmental cancer, 67 percent scored Bruce Ames high, 19 percent medium, 6 percent low, and only 8 percent did not know his work. Sidney Wolfe, a critic of the modern American diet who is frequently on the national media as an expert, was only rated high by 24 percent, medium by 15 percent, and low by 11 percent, with 50 percent not even knowing who he was. Similar numbers, 24 percent high, 20 percent medium, 17 percent low, and 40 percent don't know, were attained by Samuel Epstein. Epstein, a firm believer that "chemicals" in the environment are the leading cause of cancer, is often cited in books and the media as a leading, if not the leading, expert on cancer. The survey findings on what substances are highly carcinogenic and which are not would also come as a great shock to those who believe that chemicals are killing us (Rothman and Lichter 1996, 231–245). In fact, there have been innumerable surveys of the public and experts on what is dangerous and what isn't, and inevitably the technological is always considered more dangerous, frequently much more dangerous, than the judgment of expert opinion and a study of the actual mortality rates and their causes would indicate.

Each year at the Thanksgiving/Christmas holiday season, the American Council on Science and Health (ACSH) issues the usual holiday menu with a detailed list of the many carcinogenic chemicals each item contains. They also advise the reader to enjoy their holiday meal, as the carcinogens are in very small amounts. The point is that the dose is important and that there are natural carcinogens in greater quantity in the foods we eat than in any trace of pesticides which may still be on the food. The position of Ames and ACSH is that the vast majority (99.9 percent or more) of carcinogens that we ingest are natural products of the foods we eat and not manufactured chemicals. This position is in line with two different National Academy of Sciences panels established to examine toxins and carcinogens in our diet (NRC 1996; see also NAS, 1973).

In addition to a presumed upsurge in traditional illnesses "caused" by "chemicals," we have a variety of new illnesses such as multiple chemical sensitivity. Since a "chemical free" environment is a meaningless concept, we still have to address the problem of why the sensitivity to manufactured chemicals and not to naturally occurring chemicals, including plant toxins? Further, in opposing irradiation of raw produce, we are left with only less effective chemical means of attempting to cleanse them of harmful microorganisms (Burnett and Beuchat 2001; Holliday et al. 2001). The fact is that we may have been too successful in creating a more hygienic environment leading to other problems. Good hygiene makes good sense but obsessive hygiene— "the antibacterial craze"—can be counterproductive since it is as meaningless to be free of all microorganisms, including the sometime harmful ones, as it is to be free of all "chemicals." "Some researchers have found a correlation between too much hygiene and increased allergy." Studies have "revealed an increased frequency of allergies, cases of asthma, and eczema in persons who have been raised in an environment overly protective against microorganisms." One scientist has "likened the immune system to the brain. You have to exercise it, that is, expose it to the right antigenic information so that it matures correctly. Excessive hygiene, therefore, may interfere with the normal maturation of the immune system by eliminating the stimulation by commensal microflora"(Levy 2001).

To some, our discourse on romanticized notions of the lifeways of other cultures or what is "natural" might seem to be iconoclastic. In fact, researching this volume was an iconoclastic experience for me, as it exploded many of the cherished realities of my youth. Whether deliberately or unintentionally iconoclastic, I am unashamed, and unapologetically so. Being an optimist and a believer in ever expanding human potential, I would much prefer to research and write about humankind's triumphs and great achievements. But when some groups falsify and romanticize the past of other groups in the service of an ideological agenda, then looking at the negatives is necessary to counter the mythology and to establish a more realistic knowledge basis for policy formulation. This look at the dark side of the lives of others and earlier peoples is not a denial of the great contributions peoples of all cultures and throughout human history have made to art, literature, music, science, and technology in the face of what would seem to us today to be unbearable hardships. It is only fair to point out in this context, that it is the antimodernists who appear to take pleasure in every failure in modern life, as it seems to verify their ideological preconceptions about the destructive power of science and technology. Not only have NGOs grown and prospered by finding fault with modern life, but entire multi-billion dollar industries in organic food and alternative medicine have as their basis the dangerous shortcomings, if not life-threatening conditions, of modern agriculture, food production, and medicine.

One should, in fact, having nothing but the highest admiration for a series of stone age hunting and gathering technologies that sustained humankind through 99 percent of our existence. We sometimes marvel at the ingenuity of these people as they solved the problems of life that they faced with such limited technological means. Technology is a dynamic, cumulative, accelerating process as each new tool or technology, new ideas or knowledge, or new skill or capability adds to the base from which the process was furthered by combination and recombination of these elements. The smaller the existing base of capability, the less opportunity for new combinations and the slower the potential for change. Our ancestors endured and persisted and laid the foundations for the progress that we now enjoy. Knowing nothing different, they endured and found meaning, and even joy, in conditions of high mortality and short life spans that we with our heritage would find unbearable. And that is simply the point: However great our admiration for their endurance and however great our gratitude for their role in furthering the process that now allows the privileged lives that we lead, their lifeways are simply not an option for us today in any respect, and characterizing them as other than what they are is counterproductive to creating a satisfactory life for those alive today and for those who will come after us. Even if for some reason we wanted to return to an earlier technology, continued survival requires continued technological change. We humans adapt our lives to the cumulative technologies of our time, so any attempt at return would take an enormous toll in human lives. As Veblen perceptively noted, "here and now, as always and everywhere, invention is the mother of necessity" (Veblen 1922, 314; see also Hamilton 2001, 745–747).

Romanticizing the lives of the poor contributes nothing to alleviating their poverty. What is needed is to protect the rights of smaller, politically and economically vulnerable groups, not their cultures. Empower people and then let *them* decide what they wish to preserve in their culture, what they wish to retain in modified form, and what they may wish to abandon.

Antitechnology romantics can be found across the political spectrum, but most congregate at the extremes in democratic societies. Some find "nature" or utopias in the past while others find "nature" in earlier technology in our culture or in the lifeways of other ethnic groups, past, or present. One author argues that immemorial traditions are beliefs that we learned in our youth.

> Each generation sees its culture as that with which it grew up. Its hallowed values and traditions are those it learned in childhood. Many elements that it values as its culture were controversial foreign imports a generation or two ago (de Sola Pool 1979, 145–146).

The Ready-made World

We can all agree that the natural beauty of the world and the wonders of the universe preceded the emergence of humanity. However, as our narrative seeks to make clear, humans were not preprogrammed to have some innate aesthetic appreciation of this beauty, and it is human inquiry and knowledge and the other rudiments of civilization that have revealed these wonders for us to behold and that drive us forward to learn more and to create our own wonders. It can be considered among the highest achievements of humanity to recognize this marvelous civilizational heritage and have a desire to maintain a stewardship of it as well as respecting the rights of other creatures' rightful place in sharing it with us. In the very real sense then, humans are cocreators of "nature" and it makes sense for us to preserve and protect what we have participated in creating. However, to preserve and protect is not and cannot be an encapsulation of a static entity, but one of fostering a process of change in an intelligent and sustainable manner. What we need to preserve above all is a human sensibility about life, our lives, and the role that we play in the continually emerging larger scheme of things.

Once we believed that truth is beauty and beauty is truth. Knowledge, be it scientific or technical, is every bit the pursuit of truth and beauty as is any other form of inquiry. The beauty of the "nature" that life before us, and our understanding of it, has created is revealed to us in many ways that have equal claim to legitimacy. All technological and scientific inquiries are simply different ways of accessing and understanding the world—nature—around us. To those poets who claim that science somehow "takes away from the beauty of the stars," the physicist Richard Feynman counters that he too sees the stars and feels them, allowing the "vastness of the heavens" to stretch his imagination. Speaking of that "vast pattern, of which he is a part," Feynman adds it does not do "harm to the mystery to know a little more about it. For far more marvelous is the truth than any artists of the past imagined" (Baeyer 2000, 14).

Feynman asks a fundamental question—why can't the poets find beauty in scientific truths such as physics and astronomy?

Why do the poets of the present not speak of it? Why men are poets who can speak of Jupiter if he were a man but if he is an immense sphere of methane and ammonia must be silent (Baeyer 2000, 14).

We need to deconstruct and identify the fundamental assumptions and differences in the narratives that inform the competing discourses on "nature" and the "natural." "All natural" is superior to any human creation only if the "natural" somehow exists or existed apart from any human intervention. Nature produces it, and humans harvest it. Or at least, any productive process, particularly in food production, is superior to the extent that it uses "nature" and follows its laws. Proximity to nature makes us healthier and happier, or so some would have us believe.

At the same time that we humans are expected to be as in tune and in harmony with "nature," there is an implicit belief that "nature" or at least some segments of it—wildlife parks and wilderness areas—are better off without us or at least more authentically "natural" without us. "Living lightly" on the land implies that there is a pre-existing habitable land that humans occupy, and "living in harmony" with nature equally implies a pre-existing benign, provident "nature" awaiting our emergence or arrival. We have been told repeatedly over the last three decades that we did not inherit the earth from those who came before us but are simply holding it in trust for those who come after us and have a sacred obligation not to diminish it. Whether it is a theological belief or an evolutionary one, the belief is that we humans emerged in a world that was a ready-made cornucopia for our sustenance. Our task is simply to find a way to live within nature's limits and in terms of its laws. To many, but not all, holding this view, "technology" is the great destroyer. Ironically, it is those of us who believe in the creative potential of science and technology to sustain us who are deemed to be cornucopian, and not those who believe in a pre-existing cornucopia.

Jeremy Rifkin presents an extreme version of this thesis using the concept of entropy (Rifkin et al. 1980; see Georgescu-Roegen 1971 for a more thoughtful use of the entropy concept). The earth began with low entropy—differentiation and the capacity for work and change and therefore a good condition—and ineluctably moves towards higher entropy—uniformity and diminished capacity for work and change. Taken literally, Rifkin's thesis would have us living frugally (using renewable resources, recycling, etc.) but not sustainably because thermodynamic doom can be delayed but not indefinitely postponed. The continuing popularity of Rifkin's "new world view" is difficult to fathom, as his use of entropy would apply to a closed system. Whether Rifkin or his followers have noticed or not, the earth is an open system with the sun providing a stream of energy allowing for the development of complexity and the emergence and evolution of life.

Our view is almost exactly contrary to that of Rifkin. Earth was neither ready-made for life nor was it later ready-made for humans. Earth was obviously ready-makeable for life and for humans, which is true by definition, since life emerged and we are here. Some of the most fascinating writing in contemporary popular science discusses all the special properties of matter and energy that made the larger cosmos possible. If conditions had been slightly different, we would have had an unrecognizably different universe or Earth that could not have supported life as we know it. Equally fascinating are the special properties of our solar system: The size of the sun, our distance from it, our moon as a stabilizing force for the Earth's rotation, and Jupiter's role in drawing in large objects that if unimpeded could have struck the Earth with devastating consequences, all of which set up conditions that encouraged the development of life on Earth. Whether any other "universes" (a contradiction in terms?) existed before ours, exist now, or are yet to come, with different properties, or whether there are other planets as favored as ours, are issues for specialists to debate.

The Earth's first life forms were heterotrophic, which means they could not manufacture their own nutrients and therefore subsisted on a cumulated store of organic compounds that had been created and was continuing to be created by abiotic processes such as lightning discharges in the earth's atmosphere, or by the sun's ultraviolet light under conditions in which there was not yet an ozone layer shielding the earth from their intensity. In this sense the earth was ready-made for life, but for a very limited form of it. The organic compounds were converted into energy for these anaerobic unicellular organisms by chemical fermentation. Oxygen was in every way a dangerous pollutant for these early life forms (DeGregori 1985, 5).

As life expanded it was using more organic material than was being created; pre-existing stocks were being depleted. Evolution solved part of this energy crisis by photosynthesizing organisms that could use the sun's radiant energy to convert carbon dioxide and water into glucose and oxygen. But there was still a need for nitrogen, which is present in all known life forms in amino acids. "At some point the demand for fixed nitrogen exceeded the supply from abiotic sources," creating a "possible nitrogen crisis for Archaean life." As long as adequate sources of nitrogen were available for early life, there would be no survival advantage if the ability for nitrogen fixation emerged, since "biological nitrogen fixation is energetically expensive." But when demand for nitrogen exceeded the abiotic supply, whether from depletion of prebiotic sources or from the emergence of higher plants, the development of energetically expensive "metabolic pathways to fix nitrogen" could have a survival value if it arose, which it did (Navarro-Gonzalez et al. 2001; see also Heckman et al. 2001).

Fortunately for us, life evolved energetically expensive metabolic pathways to fix nitrogen, allowing life to continue to evolve higher forms. We do not know whether similar early heterotrophic life forms emerged on other planets where the chance evolutionary development of photosynthesis and nitrogen fixation did not occur, thus causing the emergent life to be extinguished. What we do know, is that it was the evolutionary processes of life itself which the made earth habitable for a wider variety of life forms including ourselves. In these emerging higher life forms, there were new uses for nitrogen. "Although relatively scarce, nitrogen is present in every living cell" (Smil 2001, xiii).

If plant life, securing their energy needs from the sun by photosynthesis, had not emerged, the pre-existing organic energy sources would have been exhausted. But life forms still needed help in obtaining their nitrogen needs. The atmospheric N_2 molecules "must be split into two constituent atoms before they can be incorporated into an enormous variety of organic and inorganic compounds" and for this they needed the help of another organism, nitrogen-fixing bacteria. "There is only one group of living organisms capable of Nitrogen fixation, about 100 bacterial genera, most notably *Rhizobium* bacteria associated with the roots of leguminous plants" (Smil 2001, xiv). Life on Earth faced a potential nitrogen crisis. Nitrogen is abundant in the Earth's atmosphere but it is not in a form (called "fixed") life can use to create amino acids which in turn form the basis for proteins.

> Nitrogen is an essential element for life and is often the limiting nutrient for terrestrial ecosystems. As most nitrogen is locked in the kinetically stable form, N_2, in the earth's atmosphere, processes that can fix N_2 into biologically available forms—such as nitrate and ammonia—control the supply of nitrogen for organisms. On the early earth, nitrogen is thought to have been fixed abiotically, as nitric oxide formed during lightning discharge (Navarro-Gonzalez et al. 2001).

There are some lessons here applicable to human resource issues. Nitrogen in a form usable by the existing heterotrophic life was a fixed finite resource and as such was inherently exhaustible even though in this case it was being renewed by abiotic sources. "Living within limits" would have meant, at best, no evolution to higher forms, though one has difficulty imagining the mechanism for a continued expansion of life in some form to the limits of resource availability without increased death being the force that kept resource supply and demand in balance. Though nitrogen is essential, it is not a resource unless it is in a usable form. Beyond the fixed nitrogen created by lightning, atmospheric nitrogen, roughly 80 percent of the Earth's atmosphere by volume, *became* a resource when life evolved the means of using it. "Resources are not; they become" (Zimmermann 1951, 15).

More relevant to our inquiry is the Earth that existed at the time of the evolution to *Homo sapiens*. For early proto-hominids, there were limits as to which environments could be inhabited. The evolution of knowledge, intelligence, skills, and tool-using allowed our earliest ancestors to move out of their original tropical habitat, becoming the only mammal to be able to inhabit the entire globe without speciation. Rather than the slow process of biological evolution allowing the organism to settle in new territory, humans used technology to make new environments hospitable to them.

Everywhere the humans went, technology was used to make the environment habitable. For example, the far northern cultures of Europe are among the richest and most developed regions of the world today. If we give the analysis of these areas some historical depth, we will find that some of the colder areas of Europe were first settled by the Neanderthals, who, it is believed, were biologically more adapted to cold weather. They in turn were displaced by *Homo sapiens* which, modern evidence seems to indicate, had a better developed hand, wrist, and arm for tool using and tool making and possibly a greater capacity for culture (Niewoehner 2001; Churchill 2001; Balter 2001). Biology that facilitated technology trumped biology alone.

Humans could not go very far north in Europe beyond the line where animals either hibernated or migrated in winter, until they developed the ability to engage in various forms of communication, as well as abstract symbolic thinking, which in time allowed them to think and use their knowledge to intercept migrating herds in order to obtain the essential nutrients that could only be derived from killing and eating animals.

Though recent research has placed tool-using primates (Neanderthals or *Homo sapiens*?) in the Arctic region of Russia earlier than previously thought, it is still believed that advanced mental and technological abilities were required to survive there (Pavlov et al. 2001; Gowlett 2001; Wilford 2001).

> We believe that survival of humans in this arctic environment on a year-round basis would have required long-term planning and an extended social network, qualities that are generally associated with modern human behavior (Pavlov et al. 2001).

For humans, any given level of technology defines a set of useable resources that are fixed, finite, and inherently exhaustible. The question is not whether the potential for a resource crisis is an integral part of the human endeavor, but how we should respond to it. As we discussed in an earlier chapter, we hominids have been hunters and gatherers for over 99 percent of the time we have existed (Lee and DeVore, eds. 1968). Human ideas and skills took the raw stuff of the universe, stones of certain kind, and turned them into tools, which enhanced the resource character of the environment by allowing for a more intensive exploitation of the environment for food and for expansion into new environments. Population growth eventually meant that hunters and gatherers reached the limits of harvestable resources with their complement of technology.

However slowly population was increasing, it was growing faster than technology was expanding the resource base, creating a crisis. "Living within limits" to keep the human population to a few million of us would have required increasing the already high death rates, shortening the already low life expectancies. The response to the resource crisis was technological change in the form of the development of agriculture.

Agricultural history since these early beginnings was one of human innovation which transformed or amended soils in some way as to make them arable, and of the adaptation of plants and cropping methods to allow increased output or the spread of agriculture into new areas. Every item in this process, the domesticated plants and animals, the land and so on, became resources because human intelligence developed technologies to allow their use or their more intensive use. The same could be said for every other artifact of human life. Minerals became resources when humans could transform them into metals. "Resources are not; they become." Technology creates resources and if we use them creatively to advance science and technology, then we can create resources faster than we use them, expanding and not contracting the resource base, making them less scarce, not more so. This is what humans have done for much of our history and what we did throughout the 20th century, making possible its extraordinary achievements for expanding human life.

Synthetic Nitrogen and Agriculture

In 1828 chemistry professor Friedrich Woehler (1800–1882) accomplished the first laboratory synthesis of an organic compound (specifically, urea). He thus proved that chemistry could duplicate, even without organic molecules, a product of animal metabolism. The vitalists of Woehler's time maintained that organic molecules could not be derived from inorganic molecules. Another chemistry professor, Justus Baron von

Liebig (1803–1873), a cofounder of agricultural chemistry, refuted the theory, then prevalent, that only organic material (specifically, humus) nourished plants. Among Liebig's highest achievements was his discovery that minerals alone could fertilize soil. Wide acceptance of this discovery has enabled better nourishment of humankind—despite humans today numbering more than six times what they were before the discovery. In 1845 one of Woehler's students, Adolph Wilhelm Hermann Kolbe (1818–1884), accomplished the first synthesis of an organic compound (acetic acid) from its elements. To Sidney Toby (2000), "the death-knell of vitalism in chemistry was sounded."

An increasing scientific understanding of the world was simply unacceptable to the true Romantics. Organic chemistry may have sounded death-knell of vitalism, but the Romantics refused to hear it. For some, the triumph of chemistry made a focus on vitalism even more imperative. The much maligned "reductionist" science of chemistry which by use of analysis was able to synthesize organic compounds in the 19th century, was able, in the 20th century, to perform what may be the most important synthesis of the century, the synthesis of ammonia and its use in the creation of urea for agriculture. The continued 19th century advance of chemistry in which German scientists played a leading role, laid the foundation for the early 20th century work of Fritz Haber and Carl Bosch in the industrial synthesis of ammonia from atmospheric nitrogen, allowing for the mass production of synthetic nitrogenous fertilizer (Smil 2001). Humans could now take this most abundant of the atmospheric elements and convert it—manufacture it—into the most vital resource for the growing of crops and the creation of nutrients to feed humans.

At the time of this vital development, a limits-to-growth theorist could have raised serious questions not only as to whether the growth in food supply could continue but even whether it could be sustained at its then current level. For agriculture, the potentially limiting nutrient was nitrogen in a form that was usable by plants. Europeans and North Americans were literally mining guano and nitrates in the rest of the world to provide nutrients for their agriculture and food production. These were clearly exhaustible resources that were becoming increasingly scarce. The frontier in the United States had officially been declared closed, so a new nation was now complete, but so were most of the gains from simply bringing new lands under cultivation in the United States and most everywhere else.

Before modern chemical pesticides were an issue, the foundation of organic agriculture for Rudolf Steiner was opposition to synthetic fertilizers, since they were "man-made" and alien to the environment, and most of all because they were "dead" (Bramwell 1989, 20; also, Steiner 1958; Ferguson 1997). The "vitalist" reaction to organic chemistry was almost immediately applied to synthetic fertilizer, and the food thereby produced was attacked as lacking some vital life force. The "vitalist" attack against modern science and its fruits continues to the present, as does the skepticism about the "project of mastering nature." And as Rifkin argues and advocates, it comes from both the right and the left of the political spectrum (Rifkin 2001; *Economist* 2001b; Postrel 2001).

Never the Twain Shall Meet?

In the discourse and disagreements over issues of the environment, conservation of wildlife and habitat, energy and agriculture, extreme polarization is too often the norm. Describing differences between various participants in terms of the dichotomies described previously is not meant to contribute to further polarization. Rather, we offer them as ends of a continuum with various groups tending towards one end or the other. As we have already noted, New Agers, romantics, and many conservationists implicitly assume a benign, somewhat cornucopian nature that must be protected from man and technology. In other words, there is a belief in the ready-made world that is in danger of being destroyed by humans. It is also a natural world whose "natural" products are endowed with a not readily definable or verifiable goodness or superiority. In foodstuffs it is very clear: That which is "natural" embodies vitalist, life-giving properties. Whether formally stated or not, vitalism is at the heart of popular fears about genetically modified food.

Limits-to-growth was a consistent and central theme of these same groups in the 1970s and early 1980s as the threat of resource exhaustion gave rise to a demand for renewable resources and the need to live frugally. Frightening scenarios about calamities from population growth generally assumed some fixed quantity of "natural" resources in danger of being exhausted, i.e., limits-to-growth, and were particularly strident in the 1960s. Declining population growth rates and low commodity prices have subdued the limits-to-growth rhetoric in recent years but it remains implicit in most of the antitechnology narrative.

The Alternative Narrative Framework

Our task has been to present an alternative set of explanatory propositions to the ready-made world, limits-to-growth and vitalism. More important, since we argue that not only are these propositions erroneous, then any policy prescriptions derived from them will likely be flawed in some way. Our further task, then, is to provide a framework as a basis for devising operational theories leading to policies that achieve the broader objectives of conservation and sustainability.

Some elements of the environmental policy agenda do emerge from our study. In the conservation of wildlife, one main element of any policy is obvious and uncontested but not as easily implemented. Every time one criticizes wildlife policies and practices, as I did previously, its defenders claim that these are outdated policies and that the current mode is one of conservation to the benefit of, and involvement by, the local population. Strange as it may be, there seems to be broad agreement on what the policies should be, but some of us have trouble finding cases where they are actually being carried out as claimed. What is needed, then, is to continue to publicize the injustices and place unceasing maximum pressure on governments and NGOs to live up to the standards they proclaim. In giving local people the choice of living in and around the preserve and an opportunity to profit from wildlife conservation, we are protecting cultural diversity as well as biodiversity. Cultural diversity resulting from local economic opportunity is not one of seeking to preserve a nonexistent pristine untouched habitat or culture, but one of allowing peoples and their habitat to evolve in an orderly manner of their choosing.

Conservation of habitat requires a high technology strategy particularly in the form of intensive, high yielding agriculture. Whether or not one disputes, as I do, the claims of "organic" agriculture about being "environmentally friendly," the simple fact remains that the less intensive the agriculture, the more land that will need to be brought under cultivation, and there will be less left for other flora and fauna. This also means that more marginal lands have to be cultivated with inevitable environmental degradation. Modern biotechnology has the potential for bringing previously degraded lands back into cultivation with, for example, salt tolerant plants that could be cultivated on lands salinated by centuries of irrigation. This would also relieve or reduce pressure to bring other lands under cultivation. If we are to fulfill the promise of modern technology and agronomy, then we will not only have to find ways of feeding those among us who are still hungry, but also feeding an additional three billion that are expected to bring global population to nine billion by the middle of the century before leveling off or even declining.

The hysteria over population growth has subsided but the issue remains important. In an earlier book, I was highly critical of those who sought to control population through what I would call a death strategy (DeGregori 1985). Triage policies were advocated which would have written off entire countries and populations as not able to be saved. The lifeboat metaphor was used with the suggestion that trying to save everyone would result in dooming everyone in the overcrowded lifeboat. One author opposed any U.S. policy of exporting "death control" capabilities, considering it to be a prescription to prolong misery and threaten the planet. The very clear implication of the triage proponents was that it would be better for all if the wretched of the Earth die sooner rather than later.

I propose a life strategy for eventually controlling the population. As we have previously seen, even rates of population growth that remained very low because high mortality rates eventually lead to environmental stress absent any enabling technological change. The rapid population growth of the last century has been largely if not entirely the result of falling death rates. In fact, for the last half century birth rates have been falling more rapidly than at any time in human history. For the first decades of the last half century, death rates were falling faster than birth rates, giving rise to a seemingly unsustainable accelerating rate of population growth. Yet some of the areas like East Asia that were experiencing the highest rates of population growth in the 1950s and 1960s were those that were initiating economic growth and rapidly bringing down death rates. This in turn would eventually lead to falling birth rates and low fertility rates. This should have been expected, since in previous centuries it was those countries at the forefront of scientific, technological, and industrial development that had the most rapid rates of mortality decline and population growth. Many of the European countries that once led the world in population growth now define the "population problem" as one of keeping population from declining.

The last quarter century has seen birth rates falling faster than death rates with the population growth rate slowly but surely declining and expected to reach zero (or below) by mid-century if not before, but with a population in the vicinity of nine billion. The last quarter century's slowing of population growth just about offset the earlier acceleration.

Some figures that I recently cited bear repeating. If the high birth *and* death rates of 1950 had prevailed to the present, then world population would be about the same as it is today (actually it would be a few hundred million larger). But there would be many more births and many more deaths as a result of the higher mortality rates. However, the continued forward projection of the 1950 birth/death rates would result in a global population of 15 million and growing, instead of the expected circa nine billion (Heuveland 1999, 690). So we can argue that a life strategy is not only morally and ethically superior, it is more effective at stabilizing population with longevity and good health.

A life strategy for population management would entail:

1. Economic growth and development strategies that focused on education, particularly of women, child survival, and other broadbased measures that slow down mortality rates while raising incomes. Intensive agriculture will continue to be necessary for any nutrition component of a child survival program. Reducing infant and child mortality rates does initially increase population growth rates (almost by definition), but it is essential for reducing fertility and population growth rates in the long run.

2. Choice is essential. In the political discourse in the United States, pro-choice and pro-life are seen as opposing positions. Thus, to avoid this controversy, we simply speak of strategies of choice and life. It has become increasingly clear that if women have: (a) the education to enable them to make informed choices, (b) the access to child health and survival care for the children that they have, (c) the income to afford contraception and access to it, and very importantly, (d) the freedom within their society and spousal relationships to have the right to choose, then enough women will choose to have fewer children to bring the fertility down so as to bring population growth rates under control. Freedom to choose means the freedom to have children as well as the freedom not to have children, or to limit the number one has. Those who in the past advocated forced sterilization policies were as wrong about the dynamics of population as they were about fundamental human rights. Population hysterics have always favored "family planning" programs, but the lessons of the past decades show that their effectiveness is a function of their being part of a larger development process.

3. Resource creation through technological change is key to overcoming any presumed limits-to-growth. A resource-creating economy is a knowledge-based society with respect for knowledge and those who have it on the basis of what scientific and technological inquiry has already brought us. Not only have we been creating resources faster than we have been using them, but the raw material—"natural resources"—proportion of the economy has a long history of being a declining percentage of economic output as the economy expands. More recently, as many researchers have noted, we have been "dematerializing" our economy as greater efficiency in communications

(fiber optics), or in the strength of materials, has meant that we are actually using fewer material resources by weight as the economy expands (Scarlett 2000; Bailey 2001). Blaming modern science for all of our problems merely compounds them and robs us of the most effective means of solving them. The considered judgment of experts doesn't close out inquiry on any issue, be it global warming or genetically modified food, but intelligent and effective inquiry and problem solving policy formulation is impossible without expert input.

4. Much of the current talk of education reform has been focused on basic skills, such as reading and writing. It is undeniable that literacy is fundamental. However, there is also a need for scientific and technological literacy, and more effort needs to be devoted to public understanding of scientific issues. Both those who sincerely believe that "chemicals" are killing us and those of us who don't should welcome a greater public understanding of the science, or alleged science, that underlies these issues. Some knowledge of statistics is essential, including such basic propositions that random does not mean uniform. Consequently, a cluster of some human malady need not have a *local* cause, and even if it does, it is not necessarily an errant man-made chemical. Science and the philosophy of science teach us that all effects have a cause, but statistics teaches us that it need not be a local cause.

5. What we are advocating is the demystification of public policy formulation. Those who find a certain richness and profound satisfaction in the mysteries of life have a plethora of activities—love, art, music, etc.—in which to find ineffable meaning, vitalist's essences, and unverifiable truths without the need to confuse pubic policy discourse with them. We need evidence-based public policy formulation. If "all natural" is better, then we need an educated public to demand the evidence for it or against it. In other words, there should be no privileged or imputed a priori superiority to any position except for that which has consistently been based on evidence. No groups, NGOs or others, should have some special claim to be able to divine the dangers of that which is unknown—ignorance is not privileged, nor do we need an NGO priesthood to define danger where there is no evidence for it. Vitalism from the origins of organic chemistry found vital, unmeasurable, and unobservable essences in that which was natural and organic. What we might call neo-vitalism carries these beliefs forward and adds equally unmeasurable and unobservable dangers in that which they pro-

claim to be contrary to nature. However clever the metaphors may be that "chemicals" or inserted genes in our food are ticking time bombs, they should not be allowed to drive regulatory policy *without supporting evidence*, which so far has not been forthcoming. Added to a basic knowledge of statistics is the necessity for some public understanding of risk analysis and an understanding that totally risk-free alternatives in life simply do not exist. The belief that all problems have a solution must be matched with the equally important proposition that all solutions have problems. Our policies to solve these problems must be science-based both in analysis and proffered solutions in an ongoing process of intelligent evidence-based change.

6. Earth was not ready-made for life but, by definition, it was readily adapted for initiating it. The habitats within which humans evolved were not ready-made for human life as we know, but we did make it so through technological change. It is ours to continually make and remake. *Neither the criticism of the critics nor the claim for the enormous benefits of technology and science should be interpreted by the reader as a denial that a world of 6 billion humans and a yet-to-be world of 9 billion humans is not confronted with a staggering array of environmental and other problems.* What we argue is that the romantic/vitalist/antitechnology understandings and advocacy are prescriptions to worsen our problems, not improve upon them. Science and technology might not alone save us, but we cannot save ourselves without them.

CHAPTER 9

Technology and the Promise of Modernity

Many in modern society are experiencing what the classical scholar, Gilbert Murray, called a "failure of nerve" (Murray 1951, Chapter 4). Making a comparison to Germany in the 1930s, Pois maintains that "people fear change, other than those technological advances perceived of as being immediately beneficial, and American politics is infused by a longing for timeless stasis" (Pois 1986, 153). Anti-modernism comes in many forms, but the ultimate (or close to it) would be exemplified by the author, John Zerzan, who argues that humankind made a "monstrously wrong turn" with the development of speech and symbolic thinking (Zerzan 1998, 260–263, 268, 273). It is not at all clear how one would enter into an honest dialog with Zerzan or his myriad of followers, since we presumably would have to use speech and symbolic thinking to do so. In any case, his merry band of followers have become major participants in street demonstrations and remain true to their beliefs, noticeably lacking in intelligent speech and symbolic thinking.

To Rene Dubos, "like it or not, from the moment we learned to transform things according to functions we developed a hundred thousand years ago, we drove the natural out," which to Zerzan would further compound the error of our ways (Dubos and Escande 1980, 99). Dubos uses the term "Dr. Pangloss" to characterize the nature-knows-best school devotees. Voltaire's Dr. Pangloss (in *Candide*) thought we lived in "the best of all possible worlds" (a phrase that originated in the writing of G. W. Leibnitz). The irony of this designation is that critics of modern technology frequently refer to its supporters as being Panglossian.

Since World War II, a number of writers equated the Nazis and the Final Solution with the Enlightenment, instrumental rationality, and science. Given the horrors of the Nazis and the Holocaust, other criticisms of modernism/science/technology may seem tame by comparison, but still many postmodernists and "eco-feminists" in one form or another equate modern science and technology with violence, oppression of women and minorities, environmental destruction, and ultimately with death. This returns us to a slight variant of the question that I raised in the Introduction: If everything in modern life is killing us, why are we living so long?

In the 1970s, the antimodernist rage was the *Limits to Growth* book (Meadows et al. 1972) and the belief that through some "small is beautiful" use of renewable resources strategy we could learn to "live within limits." This thesis went hand-in-hand with the population alarms that were forecasting mass famine and death. Time does move on, and those with the most frightening prophesies of impending calamity will find that the day of forecast doom has arrived but the prophesied catastrophes have not happened. In fact, all the evidence indicates almost the exact opposite. Some try to compound earlier errors by attempts at denial, claiming, for example, that their doomsday prophesies were only scenarios, while at the same time trying to claim credit for favorable changes (declining fertility rates) that allegedly followed from them. Many who come to believe in an impending doomsday will continue to have faith in those who predict it, no matter how wrong they turn out to be. One erroneous prophet was described as "brilliantly perceptive and predictive" but also having got "very little" wrong except that "he approached his subject too gingerly" (Tobias 1998, 49).

Nearly three decades after the publication of *Limits to Growth*, it is abundantly clear that we are not experiencing the resource scarcity that should have been observable by now. In fact, for most raw commodities, the problem for the producer is over capacity and low prices rather than the astronomical prices that would be expected to prevail if resource exhaustion were rapidly approaching. A process of denial by the original authors and others has been under way for some time, with the latest being a work which attempts to explain away the earlier prophesies (Meadows et al. 1992; Hawken, Lovins and Lovins 1999, 145–146, 331–332; for a critical review, see Smil 2000). Even if one accepted the supposedly "technical" arguments offered by Hawken and the Lovins as to why the original study was not flawed, it is clear that following the report, people like Amory Lovins were acting in terms of the interpretation of *Limits to Growth* that was criticized by economists and massively demonstrated by subsequent events to be wrong (DeGregori 1985a, 1985b, 1986, 1987a, 1987b, 1995, 1996). Without an interpretation that resources were "finite," there would have been no reason for Lovins and others to argue that we needed to make a transition from non-renewable to renewable resources. As I have argued a number of times, it would have vastly worsened the resource situation had this advice been more often heeded, since it became quickly apparent that there are no shortages of the non-renewable resources, but that there are legitimate concerns about renewable biological resources. To repeat, ideas have consequence and the consequences of the limits-to-growth theory was wasted investment in commodity production on the expectation of their scarcity (and therefore higher price), which helped to produce the commodity glut and the falling prices of resources in the 1980s and 1990s, benefiting affluent consumers and harming producers in low income countries.

The promise of modernity is not based on the negatives of any previous condition of humankind but on what is the open-ended array of possibilities that it offers us. Each of us has our own definition of modernity and its promise. Mine would be the belief that if every problem is not solvable, it is at least worth the effort of trying to solve it. This belief is not blind faith but is based on the observation of the great strides that we have made on so many fronts, from infectious diseases to being able to provide adequate nutrition for everyone. It is the greatest tribute to modernity when any institution falls short of what is possible in these endeavors; it is the measuring rod of modernity that is used to criticize it. And this is as it should be, for the technology of what is possible is the goal towards which we should always strive.

Tragically, those who would call themselves "humanists" are most critical of the attempts of modernity to triumph from the scourges that have ravaged humankind. These "humanists" consider such efforts to be a manifestation of hubris, and they see any failure—a disease developing a resistance to antibiotics—as some vindication of the power of nature and the futility of attempting to defeat it rather than living by its supposed dictums. Modernity is an ongoing process, and occasional setbacks are as much a part of the process as are the more frequent successes. Properly understood, even the failures provide critical knowledge for future success. Modernity is not a promise of utopia but of the possibility of continued betterment. The struggle for betterment is as much a joy, as is the delight in its attainment. Modernity simply asks the best of all of us, and its highest achievement, for which it is still woefully short, is the creation of conditions where the opportunity to strive for the best becomes everyone's birthright. The idea that everyone can have the nutrition and health care to realize their fullest physical development and lifespan, as well as the education so that they can acquire all the knowledge they can absorb and have access to a full range of aesthetic experiences, is unquestionably still a utopian dream. But what prior civilization has ever had such goals or has had even the remotest possibility of even looking down the corridors of time to see it as a future possibility? The technology of modernity has been the "technology of accessibility" and the "technology of opportunity."

Accessibility has numerous dimensions. Astronomy and cosmology have given us access to the earliest history of the cosmos, while other scientific inquiry has taken us to the sub-atomic particle. The camera and a variety of other technologies have allowed us to explore seemingly every nook and cranny on the planet and almost every facet of life on it, while similar technologies have been sent out to explore the terrain on planets, our Moon, and other moons in the solar system. As we have noted, the camera allows an increasing number of us to have a visual image of ancestors now deceased, our elders when they were younger, or ourselves as infants and children. The Internet and a variety of CD ROMs are allowing many to search farther back in their own personal and family history. Film, radio, and television archives have created a tremendous reservoir of accessible experience of the century that we have just exited.

Over a century of recording technology has allowed the preservation of sound, which has always seemed to be the most transitory of all human experience. We may have the written record of the great oratory of earlier times but we have little if any idea of what it sounded like. Archaeologists and linguists may recover evidence of earlier languages and sometimes even be able to decipher them, but do not know how they were spoken, as the sound vanished with each word uttered and died when the language was no longer spoken. Recording has not only preserved so many of the sounds of the past century and made many available to contemporary listeners, but has also expanded access to sounds, access to which was limited previously to an elite or otherwise small live audience and to those who resided in a particular place or culture.

Hans Fantel writing in the *New York Times*, recalls his childhood in Vienna when he received the Beethoven symphonies on nine "78-rpm disks, each weighing several pounds." Prior to the 1930s, Fantel's father, living in Vienna, "then at its cultural zenith ... couldn't hear Beethoven's entire symphonic legacy within a reasonable span of time. To hear Beethoven's Second, [Fantel's] father had to persuade his boss to give him two days off to attend a concert in Prague, and it took another overnight journey to Budapest to catch up with Beethoven's Eighth" (Fantel 1989). This was the Vienna where Beethoven lived and wrote his symphonies, and where they all premiered. Fantel continues:

> At a time when the phonograph was derided as "canned music" and many serious musicians considered it beneath their professional dignity to set foot in a recording studio, my father saw in the new technology something miraculous: a machine to transcend the limits of time and space that had constrained music since its beginning (Fantel 1989).

Before recording, the music "vanished" and was "always lost forever" with the end of the performance. "Now, for the first time, a performance could live on." If Fantel's father, presumably a prosperous businessman, had difficulty hearing all of Beethoven's symphonies in Vienna before the 1930s, imagine the difficulties the rest of us would have had were we his contemporary. And we must not forget that Fantel's father probably rode the train to Prague and that the information on the concert could well have been from a newspaper or magazine which received the information via the telegraph. Fantel notes that "even as a teenager" his "musical horizons" could be greater than his father's as he had "the classics on the shelf" and could even "make exciting forays into musical esoterica" (Fantel 1989). Today, those of us in the middle income bracket or higher in economically advanced countries take it for granted that we can hear all nine symphonies of Beethoven live if we so desire. And it is not unlikely that we have all nine on long playing records, cassettes, compact discs, or on all three. And it won't be just one interpretation or recording of them.

In the last two decades, with the emergence of the compact disc, there has been an extraordinary flood of re-issues of old recordings. Most are being reissued in cassette at the same time. For Beethoven symphonies, there are either six or eight complete symphonic cycles by one conductor alone, the late Herbert van Karajan. How many complete Beethoven symphonic cycles are there for us to choose from? Dozens? Hundreds? One could count, and undoubtedly someone has. It is not only in classical music but in all areas of music that long out-of-stock recordings are now being reissued. The equipment for remastering and cleaning up an old recording in many ways makes the reissue not only superior to the original but also to the live performance. There are different methods of preparing a work for reissue. Each has its strengths and weaknesses, its advocates and detractors; each has labels that use a particular method, further expanding our array of choices (Fantel 1990). As we have noted previously, the new technologies have greatly expanded our horizons, with access to an incredible diversity of musical styles and traditions. This same profusion and multiplicity of releases also allow us to explore one tradition for greater depth in comparing, for example, different interpretations of the same work. There are CD ROMs of various great works of music in which one can listen to the music and select from an incredible array of options on the monitor for learning about a work, its history, its score, various interpretations etc. etc. etc. And for that matter, the breadth and depth of understanding are not mutually exclusive. Each contributes to others with our technology, providing opportunities for aesthetic experience and understanding that are limited only by the time we have to devote to it. If we have the technology, we have the choice of pursuing our aesthetic interests.

A common complaint is that "something is lost" because of our technology. In music, it is clearly not the case. The tape recorder has helped to preserve musical and oral traditions that would have otherwise been lost. And though the living music of African pop singers gets more publicity today, it has not lessened the availability of traditional African music. One can pick up catalogs and find recordings of folk and traditional music from all over the world, including regional styles in the United States, such as Cajun or Zydeco, interest in which was previously mainly local. Many of us acquire by chance CDs or cassette tapes of music—that of the Sephardic Jews of Turkey, for example—that we previously didn't even know existed. And the list of recorded composers in the Western classical tradition grows ever larger.

What the technology of accessibility has created is popular culture. The technologies of accessibility have given rise to cultural art forms that are themselves highly accessible. To some elitists, popular culture is an oxymoron and vulgar in every sense of that term. This assumes that to be great, culture has to be esoteric and a limited good restricted to an elite for its own preservation. As such, it is the cultural counterpart of the limits-to-growth theories. Some critics have stated that when jazz was at its peak in the 1920s to the early 1960s, great music was also popular. It was popular because it was accessible. It should be noted that it was accessible to a wider public, but except for a very few critics, its greatness was not recognized until more recent decades. As new musical forms emerged in popular culture, many elitist have feared them in much the same way that new technologies such as genetically modified food continued to be feared today. People were arrested for dancing the turkey trot. Based on "scientific evidence," ragtime music would "stagnate brain cells," "wreck the nervous system," and "lower moral standards." "At its most frightening, ragtime was a national poison that threatened to spread to the rest of the Western world and doom the civilization of the white man" (Leonard 1985, 103, 107). "In the beginning jazz was widely seen as an unwanted African virus—something that gave fevers, led to sexual deviation" (Ratliff 2001). A lawsuit was filed in 1926 by the Salvation Army of Cincinnati "to prevent the construction of a jazz theater next to a shelter for girls." The argument was made that "music emanating from the theater would implant 'jazz emotions' in babies born at the home" (Cohen 1998).

> In American newspapers—the *New York Times* included ...—scientists, ministers and politicians were installed to condemn jazz. In 1926 the *New York Herald Tribune* published a 10 count "indictment" by Dr. Henry Coward, conductor of the Sheffield Musical Union in England, that included "hooting," "atavistic tendencies" and "irritating ping-pongs by banjos" (Ratliff 2001).

In the late 18th and early 19th centuries, "there were doubts of the waltz's decency." It was shocking, repulsive and "a form of low-class lewdness" (Loesser, 1954, 159). The Nazis banned jazz in Germany in 1933, and the Communists issued diatribes and opposition to jazz throughout the Soviet Union and the countries that they controlled, resulting in an order in Moscow in 1950 to confiscate all saxophones (Ryback 1990, 11, 12; Dominick 1992, 89; Closmann 1997, 37; Sullivan 1999, 208–216). Similarly, the Japanese occupying force in Indonesia during World War II banned all "foreign popular music" (Manuel 1988, 208). After the Communist takeover of all of Vietnam, rock music "was no longer performed" (Jamieson 1993, 361). That has now been at least been partially reversed in Vietnam.

Even as these wild accusations and prohibitions were being hurled against jazz, some of the leading composers such as Debussy, Stravinsky, Milhaud, Satie, Ives, and Sousa not only showed an appreciation for ragtime and later jazz but were also influenced by it in their music (Schuller 1985, 80; on folk and non-European influence on 20th century Western composers; see also Pareles 1989a, 1989b). Others,

such as Rachmaninoff, were influenced in their performances by jazz
pianists, such as Art Tatum, the master of stride piano. As Cornell West
correctly observes, jazz was originally American but has become truly
multi-ethnic (Sullivan 1999, 235).

The reverse has also been true, as Western classical music has influenced jazz and in turn Western classical, Indian classical, Indonesian
gamelan, African traditional, and many other extraordinary musical
achievements and traditions have helped to create rock music. Some of
the previously mentioned European and American composers and others had an effect upon jazz creators—Stravinsky on Stan Kenton,
Debussy and Ravel on Duke Ellington, and Schoenberg and Bartok on
Ornette Coleman (Maddocks 1989, 14–15; see also Schuller 1986,
121–124; Schuller 1989; Ross 1988). The French jazz composer and
pianist Michel Legrand studied composition under Nadia Boulanger.
And of course, we all know the rich diversity of sources that acted upon
the Beatles. In fact, any number of rock stars have based works on classical music pieces while another musician, Elvis Costello, has performed with a string quartet, undoubtedly much to the disgust of purists
(Black 1993, 52–54; Du Noyer 1993, 49–51).

It has been a 20th century phobia of intellectuals on both the Left and
the Right that modern technology was the defining ingredient of a bleak,
bland, totalitarian future. Yet in the popular struggle against Communist
regimes of Eastern Europe and the Soviet Union in the 1980s (and 1990s),
information and communication technologies played a vital role in
providing the people with an alternative to the official government pro-
nouncements in the controlled media. From the beginning of the Cold
War, Communist authorities attempted to exercise control over information technologies such as typewriters, printing presses, and later, photo-
copiers. Yet these and many other technologies were essential if their
economies were to keep up, let alone overtake, those of the Western industrial nations. In the 1980s, when Gorbachev sought to modernize the
economy of the Soviet Union, it was clear that if this was to happen there
had to be an expanded use of the technologies, such as personal computers, that would inevitably undermine the system he was trying both to
reform and maintain.

Technology is making it increasingly impossible for governments to
keep secret what is happening in their countries from their own people
and from the outside world. Once it was typewriters, mimeograph
machines, and then copiers; later it was short wave radios (or even
small transistor radios), cassettes, satellites to see in from the outside
world, fax machines, and cell phones. All or most of these are still oper-

ative in one form or another, but computers and the Internet have become an information weapon of choice, giving a "potent liberation weapon to dissidents" (Eng 1998, 20). They are weapons for both those promoting change and for the modern Luddites opposing it.

Many of the dissident Internet campaigns are based abroad, so they are safe from clampdowns, yet they penetrate borders to spread news and views that the domestic media cannot touch. Internet activists, many working like journalists in a transnational newsroom, have transformed scattered voices into global dissident movements (20).

It would be naive to argue that these information technologies exclusively serve the cause of freedom. They can also be used by authoritarian and totalitarian regimes to further their repression. Not many technologies exist for which one cannot conceive of ways to use technology to harm others. Our argument is simply, on balance, that these technologies have been far more effectively used to promote freedom than to suppress it. This is counter to the many "Brave New World" type of futuristic novels and films over the course of the 20th century that envisioned technology as a primary instrument of repression. The 20th century has seen more than its share of despicable tyrants and regimes who used the technology available to them to murder and maim their fellow human beings. But a far more consistent and essential record of the century was the way in which technology and science have been used to advance the human endeavor.

A key technological element for the liberating dimension of popular culture was the "technologies of replication," in this case sound replication such as tape recorders and cassettes. Tape recorders began to be produced in "significant quantity" in 1960 (Ryback 1990, 44). The Soviet authorities "completely failed to pay attention to such a seemingly innocent technical branch as the production of tape recorders. A demand existed and it was satisfied, and at last, when the ideological firemen discovered the catastrophic breakthrough, it was too late" (Soviet novelist Anatoli Kuznetsov, in exile, quoted by Ryback 1990, 44). Ryback adds:

Tape-recorder production gave underground singers access to increasing numbers of listeners. No longer was the music of the bards restricted to small groups of ten or twenty people who gathered in private apartments. Tapes with underground songs soon circulated by the millions (44).

As Manuel states it: "cassette technology, the most recent mass medium, may prove to be as revolutionary as radio has been" (Manuel 1988, 6). Cassette players are less expensive to own, and cassette tapes are far cheaper to produce.

> Throughout the [T]hird [W]orld, the last fifteen years have seen the flourishing of innumerable backyard cassette industries, duplicating cassettes, printing labels, and marketing "product" through local outlets with very low initial investment and operating costs. The backyard cassette industries are able to respond to diverse regional, ethnic, and class tastes in a manner which is not characteristic of record or film industries (6).

In many respects, radio and records played a similar role of responding to, and encouraging, regional and ethnic music in the 1930s, 1940s, and 1950s in the United States, while at the same time also forging a national musical identity. Currently, in developed countries, compact discs have become a medium for preserving ethnic musical traditions and bringing knowledge and enjoyment of these traditions to people who were never even previously aware of them. For developing countries, cassettes remain a force for the recording and dissemination of music. Manuel demonstrates the democratizing potential of cassette technology.

> Cassette technology, in other terms, offers the potential for diversified, democratic control of the means of musical production, and has engendered many new forms of music which have arisen as stylizations of regional folk music (6).

American popular culture is damned, even when it is credited with an important role, possibly exaggerated, in extraordinary achievement, such as the demise of Communism. Irving Kristol saw pop culture as helping to defeat Communism because of its "corrosive effect" on authoritarian systems. Though delighted with this outcome, Kristol is otherwise "unhappy that the United States has this popular culture to export" (AP 1992). Given the emergence of American pop culture as a global phenomenon, it is hard to argue that there is not at least some merit that attracts so many diverse people to listen to or to see it (On the dominance of American pop culture, see Rockwell 1994). One author more favorably disposed to rock music argued that "since 1985, rock music has provided, in both a figurative and literal sense, the soundtrack of the Gorbachev revolution" (Ryback 1990, 3). According to a Russian historian, not only did jazz and the other popular music broadcast by Voice of America (and other propaganda organizations) have a liberating impact on people in the Soviet Union, but also those whom the Soviet government officials trained in English to counter these broadcasts became leading agitators under Gorbachev supporting *glasnost* (Zubok 1998). And one might add that throughout the region jazz and rock continued to provide the soundtrack for what became the collapse of Communism.

"In a very real sense, the triumph of rock and roll in Eastern Europe and the Soviet Union has been the realization of a democratic process" (Ryback 1990, 233). Rock music was described in the *New York Times* by John Rockwell as being "the anthem of change—racial with the civil-rights movement, and also social, sexual, and political" (Garofalo 1992a, 32). In the United States, songs such as Curtis Mayfield's "Keep on Pushing" and "People Get Ready" promoted the struggle against racial injustice (Garofalo 1992b, 235; see also Davis 1995). The technologies of replication became the "technologies of freedom" to listen to and therefore the "technologies of liberation."

> In the People's Republic of China, a style called yaogun yinyue—which roughly translates "rock 'n' roll"—gained in currency during the growth of the pro-democracy student movement. It is considered to be opposi-tional in both its lyric content and aggressive (by Chinese standards) sound (Garofalo 1993, 27).

Cui Jian, "China's principal rock icon" is still performing, but "his profile has been lowered substantially by the times." He is too popular for the authorities to ignore completely, but his message of freedom is too powerful for them to allow him to convey it too widely (Tyler 1995; see also Jones and Hallet 1994; *Asiaweek* 1995, 1997). Cui Jian's song "Nothing to My Name" (Yi wu suo you) "became an anthem of student demonstrations at the end of 1986 and again during the Tiananmen stu-dent demonstrations of 1989." "Although Cui Jian was never directly critical of the state, his lyrics were double-edged, subverting a lot of the conventions of official ideology" (Jones and Hallet 1994, 456). Even Iran has not remained immune to the "corrosive" effect of con-temporary rock music (Anderson 2001).

When the Berlin Wall was opened at Christmas 1989, people rushed through to acquire various items of Western pop culture, including cas-settes of rock music (Stevenson 1994, 95). The Czech author, freedom fighter, and later President of Czechoslovakia (and the Czech Republic after the separation of Slovakia), Vaclav Havel, gives great credit for his people's freedom to local and Western rock and roll artists, particularly Frank Zappa, calling Zappa "one of the Gods of the Czech underground during the nineteen-seventies and eighties" (Havel 1993, 116).

> It was a era of complete isolation. Local rock musicians and audiences were hounded by the police, and for those who refused to be swept aside by persecution—who tried to remain true to a culture of their own— Western rock was far more than just a form of music (Havel 1993, 116).

Ironically, Marxists of the Frankfurt School have argued that popular culture served to "legitimize the status quo by stultifying critical consciousness" (Manuel 1993, 9). The far left and the far right have much more in common than either would ever dare to admit. Other Marxists have seen the popular folksong as a mechanism for protest and revolutionary change (Porter 1991, 122). Obviously, popular culture and music cannot be all of these at the same time.

Photography has very definitely been a popular technology of opportunity and replication. It gives people a precious record of their own past and can be the means for a journey into other cultures and places. Until recent decades, when it finally became recognized as an art form, photography has been the object of an antitechnology bias, as Theodor Adorno makes clear also in his derisory reference to "mechanized art commodity—above all photography" (Adorno 1973, 5). The elitism about photography's "mechanical" reproduction is ironic because it has become one of the most humanistic and democratic technologies in modern life. It is said that only after the advent of photography (particularly after cheap, mass-produced cameras) did the bulk of the population know what their ancestors looked like. For centuries, a minuscule elite had portraits of their progenitors. Most of us were pleasantly surprised when we first encountered pictures of our parents taken before we were born or saw our own baby pictures. Though most of us take the family picture album for granted, as we do with so much of modern technology and its benefits, their loss through fire or flood is irreplaceable. During the coverage of floods (or fires, earthquakes and other events destructive of peoples' homes), the media is filled with pictures of people carrying out a few prized possessions before abandoning their homes to flood waters. More often than not, their arms are filled with family photographs. One radio broadcast described a police officer rowing to his flooded home and entering an upstairs window to retrieve a photo album and his wife's wedding dress. Previously, in a firestorm that destroyed homes in California, one news story on the most cherished items that different people saved, mentioned photographs and photo albums in almost every account (Klein 1993). These people would not accept the designation of photographs of loved ones who are now gone as being "mechanized art commodity."

Modernity and the Control of Cultural Resources

A central theme of this book concerns the control of resources. In this case, we are speaking of cultural resources and who controls them. Elitists consider themselves and their kind to have a natural born monopoly on the appreciation of the best in cultural achievement. Modernity's technologies of accessibility, and the popular culture it has generated, challenge the notion that the best in the human cultural endeavor is an exclusive province of a self-selected few. For the Third World equivalent of the elites in Western cultures, Orlando Patterson refers to the "traditional cultural gatekeepers," whose monopoly of things cultural is threatened by emerging global popular culture and the "new cultural forms" that are being created. Patterson rightly equates control of cultural resources with control of economic resources (Patterson 1994a, 3–4; Patterson 1994b, 104). Popular culture is not to be trusted because people cannot be trusted to make wise choices, therefore the correct choices have to be made on their behalf. As we have repeatedly stated, myth believed as fact and used as a basis for policy requires some form of repression to enforce it.

The real issue, as C. E. Ayres so cogently argued, was not quality but snobbery. "The presumption that what is mass-produced must therefore be inferior is itself a form of snobbery; it assumes, as a major premise, that whatever is enjoyed by many people and is perhaps accessible to all, thereby necessarily loses its distinctive excellence ... To deny the possibility that excellence could be abundant is to deny the objective reality of excellence." Ayres goes on to add that the "rejection of such snobbery does not prove that excellence is real, let alone abundant. But it does leave the possibility open" (Ayres 1961, 240). Elitism and snobbery are self-defeating and bring harm to a society at every level. "The society which scorns excellence in plumbing because plumbing is a humble activity and tolerates shoddiness in philosophy because it is an exalted activity will have neither good plumbing nor good philosophy. Neither its pipes nor its theories will hold water" (Gardner 1961).

Following Ayres, we can categorically reject the elitist criticism that modernity and its technologies are about anything less than excellence. The appreciation of excellence can be cultivated in the larger community and not limited to a self-anointed "chosen" few. A defense of popular culture cannot be interpreted as a denial of the excellence of those "classical" traditions in music, art, and literature which have survived the test of time in Western and other cultures. To John Dewey, a work of art was "recreated every time it is aesthetically experienced" and his conception of the "artistic act" was one of being a "revelation of possibilities hitherto unrealized" (Dewey 1934, 108; Dewey 1958, 359). Technological developments occurring after the creation of a work of performing art can substantially expand the potential experience of it as the performing artists use these advances to enhance the creative performance and achieve possibilities intended by the original creator but not achievable in his or her time or was simply not imagined by the original creator. Whatever the case, the validity of the performance should be in terms of the aesthetic possibilities that were realized rather then a slavish devotion to a presumed original intent that is not always clear. We can amend Dewey to read that a work of art was "recreated every time it is performed and aesthetically experienced."

Until the metal-braced (and later framed) piano came along, the performance of some musical compositions caused vibrations in the piano wires that could literally destroy them. The amount of tensile stress on the strings of modern pianos is on the order of 30 tons, making it unlikely that piano would be destroyed during a performance (Ehrlich 1990, 32; Ripin 1988, 2; Loesser 1954, 301–304). The production of pianos was conducive to the mass production techniques and technologies of the Industrial Revolution. The Broadwoods, as used by Beethoven, were better, cheaper, and produced by the tens of thousands (Loesser 1954, 232–236). Later, after 1860 in the United States, the quality of pianos continued to improve, the price continued to fall, and production rose to new highs (Ehrlich 1988, 56–63). The piano became a musical instrument for the middle class family. Contrary to the small-is-beautiful critics, mass technology is in fact technology for the masses.

Artificial fibers have given artists greater control over the sound obtainable from stringed instruments, particularly the guitar. The great violins from the late 17th and early 18th centuries around Cremona, Italy that are still played today, are strung with modern fibers (Kozinn

1999). Many a great violinist has been upset by critics who reviewed the violin rather than the performance. "It is not a tall story that an entire audience were convinced they were hearing the sounds of (Fritz) Kreisler's Stradivarius until he threw the trade fiddle away and produced the real thing" (Beament 1997, 89). Though not as dramatic, "in a famous blind test on BBC Radio 3, two of the world's most accomplished violinists, and one of its most respected appraisers, failed to detect a Guarneri and Stradivari from a British-made violin built a year before" (Schoenbaum 1998). Furthermore, "the two judging violinists were even allowed to play the four instruments before they were played out of sight" (Beament 1997, 90). There have been a number of blind tests of old and new violins since 1817 when there was one before "an audience selected by the French National Academy.... They have all produced results which one would expect from pure chance" (89). However great these instruments of earlier centuries may continue to be, we need not mystify them and diminish the possibility of other instruments played by superb musicians realizing similar greatness in artist achievement. Mystification of the instruments or any part of any artistic process needlessly limits the potential for creative achievement and public appreciation. One perceptive observer suggests:

> If a violin appears to have been made by a master craftsman, it will probably be played according. This is especially true if the player knows of the maker and his reputation (Shepherd 2000, 35).

Though postmodernists and many contemporary artists condemn technology, great artists of the past have used the available technology, and not only used it but often went to the very limits of that technology and attempted to go beyond it (DeGregori 1985, 70–73). As the production of the technologies of performance expand in quantity and quality, with a comparable increase in quality performers, many proponents of classical music are seeking to expand the audience for it. It is a worthy endeavor.

Living Longer, Living Better

In many respects, modern life need simply be defended with the fact that more people live longer, healthier lives than ever before, that there are no credible alternatives being offered that can do better, and that the technology and science of modernity offer the possibility of bringing this betterment to all human beings. Clearly, this would not be enough if the price of obtaining this betterment were slavery or some other form of oppression. But that has not been the case, as we argue also that the technologies of accessibility have been a liberating force for greater political freedom, cultural creativity, and innovation. Modernity does well on the issues that really matter. With good health and long life, an array of other possibilities emerge, and modernity has fostered the human spirit that has made the most of them.

A number of interesting indicators of healthier, better lives have emerged in recent years in addition to the standard and still valid measures of infant and child mortality, and life expectancy. One of the most interesting is the fact that throughout history, there has been a very strong correlation between growing taller and living longer. This is becoming increasingly obvious to anyone who has made repeated trips to areas in Asia and other regions where rapid economic development is taking place and where infant and child mortality rates are falling as life expectancies are increasing. In some cases, the changes in height are observable within families where one sees or has seen numbers of younger children that are taller that their older siblings, a frequency that could not be accounted for by random variation. The particularly dramatic increase in height in the population of China over the past two is documented in an article aptly title—"Richer and Taller: Stature and Living Standards in China, 1979–1995" (Morgan 2000). The data on the overall thesis of living longer and healthier, in developed and developing countries, are overwhelming; we will present a small sample of those data, drawing largely from the work of Robert Fogel, who has pioneered this research.

Over the last decade and a half, economist Robert Fogel has gone from looking at the nutrition and health status of slaves in the American South to more general questions of the relationship between nutrition and health and the evolving and improving human condition. Fogel argues cogently and with massive data that the epidemiology of chronic disease is not separate from that for contagious disease (Fogel 2000b, 295). "Malnutrition and trauma in utero or early childhood are transformed into organ dysfunction" in later life, though the mechanism for this is not fully understood. "What is agreed on is that the basic structure of most organs is laid down early, and it is reasonable to infer that poorly developed organs may break down earlier than well developed ones" (Fogel 2000a, 77). Fogel argues that "retarded development in utero or infancy as a result of malnutrition" has additional adverse consequences which become manifest in midlife or later and to the "early onset of degenerative disease of old age" (78). Inadequate nutrition can lead to a vast array (far too many to list here) of deleterious conditions that make the organism more susceptible to contagious diseases, to chronic illness later in life, and to shorter life expectancy. Nutrition in utero, even among children of normal birth weight, may also affect IQ, as at least one study shows that IQ (at age 7) is positively associated with birth weight (Matte et al. 2001, see also BBC 2001). The authors of this report refer to other studies and raise the possibility that:

> Although the reported effects of variation within normal birth weight on IQ are modest and of no clinical importance for individual children, they could be important at a population level because of the large proportion of children born of normal weight. In addition, these effects could shed light on links between fetal growth and brain development (Matte et al. 2001).

Fogel quotes a classroom ploy of David Landes to his "popular" (over 1,000 students) introductory economics course at Harvard: "Look to the left of you.... If it were not for the Industrial Revolution, two out of every three of you would not be alive" (Fogel 2000a, 44). Fogel adds:

> It drove home one of the great benefits of modern economic growth; the enormous increase in life expectancy during two hundred years that had been made possible by advances in scientific knowledge and by new economic and biomedical technologies associated with the Industrial Revolution (45).

Readers will be informed by Fogel's capsule economic and technological history from the development of agriculture through urbanization as he shows the acceleration of change through time. These changes have led to the "emergence of technophysio evolution" which Fogel defines as the "synergism between technological and physiological improvements" (74). He adds that "the most important aspect of technophysio evolution is the continuing conquest of malnutrition which was nearly universal three centuries ago" (75). These dietary deficiencies spared few if any, and Fogel's work in this area is second to none, particularly as he shows the changes in "dietary energy available for work after body maintenance" (76). Fogel finds that "technophysio evolution appears to account for about half of British economic growth over the past two centuries. Much of this gain was due to the improvements in human thermodynamic efficiency" (78–79).

In many respects the technophysio gains were modest in the early phases of the Industrial Revolution; some even argue that they were negative. There is no question that from the mid-19th century onward, the advancement of human well being was rapid and accelerated further during the 20th century. Fogel finds that in the U.K., U.S. and "other rich nations," that the "per capita income of the lower classes was rising more rapidly than was that of the middle or upper classes" in what Fogel refers to as the "remarkable reduction in inequality" during the 20th century (143).

Indeed, there was a larger increase in life expectancy during the past century than there was during the previous 200,000 years. If anything sets this century apart from the past, it is the huge increase in the longevity of the lower classes (143).

Fogel uses stature as a marvelous indicator of the decline in inequality. In the early 19th century, "a typical British male worker at maturity was about five inches shorter than a mature male of upper class birth," a gap that has been reduced to about an inch today (143–144). In four generations, the male population of Holland has added eight inches, going from 64 to 72 inches tall. Height is more than just being tall. Fogel shows how this increase in height correlates inversely with the risk of dying (146–148).

Variations in height and weight appear to be associated with variations in the chemical composition of the tissues that make up the vital organs, in the quality of electrical transmissions across the membranes, and in the functioning of the endocrine system and other vital systems (Fogel 2000b, 296).

To Fogel then, nutritional status would appear to be a "critical link connecting improvements in technology to improvements in human physiology" (Fogel 2000b, 296).

However bad discrimination and inequality may be today, from Fogel we learn that it was worse in the past. "Although there is a six-year gap in life expectancy between blacks and whites, half the gap that existed in 1900 has been eliminated" (Fogel 2000a, 166). Those of us who find the current gap to be wrong and intolerable cannot change it by denying the technology that has reduced it, nor can we be effective change agents in any way without understanding the past forces of favorable change so magnificently spelled out by Fogel. Even as there has been some widening of the income gap in the U.S. in recent decades, the "biomedical gap" continues to close. In his closing "afterword," Fogel cogently argues that our grandchildren will lead longer, healthier lives in a world in which this precious commodity is more equally distributed and widely available. Any careful examination of the evidence can give rise to no other conclusion. Unfortunately, though, many continue to demand evidence for the safety of modern food production and of the technologies that will feed tomorrow's world; when the evidence is forthcoming, the critics will always find reasons to reject it, indicating that there is absolutely no evidence that will satisfy them if it is at variance with their ideological preconceptions (Palevitz 2001).

My colleague in another department who said that the gains in life expectancies and infant and child mortality that we have discussed were just numbers, may, in a strange twist of irony, be correct. Some of the most important benefits of science, technology, and modernity are just statistics, while their failures are often very real. Almost any action, particularly in medicine, can cause harm. The Hippocratic oath of "first do no harm," if taken to an extreme, would result in no action being taken whatsoever. Taking a particular pharmaceutical can cause breast cancer but that same pharmaceutical is 30 times more likely to prevent death from cancer. Being immunized against disease has historically caused a reaction among some recipients, sometimes fatal. Though the claims of many of the adverse reactions have repeatedly been found to be without merit, no one will claim zero possibility of a severe if not fatal reaction for any medical/pharmaceutical intervention. But then again, in even in the worst-case scenarios, death from the disease (or diseases) against which one is being immunized may be 50, 100, or even several 1,000 times more likely than death from any possible reaction to the vaccine. The odds become more complicated and are often largely unknown when we move into the area of newly marketed pharmaceuticals. A policy of zero risk would be a policy of zero progress, and we would have remained at whatever levels of mortality, if not higher and growing, we were at when such a policy was inaugurated. It is not at all clear, if we instituted a zero risk policy today, whether we could even maintain the longer, healthier lives we currently have.

Adverse or even fatal reactions to a vaccine or pharmaceutical are very real and are often used as an argument against science, technology, and modernity. Of course the several hundred or several thousand fatalities that would have occurred without the vaccine or pharmaceutical would have been the price of not making and using the medical advance. The problem is that the victims of progress have an identity and rightly command our concern, while few of us know whether or not we would have been victims of not progressing unless we have had some specific problem in which our lives were saved or made far better by a recent medical advance. As long as real progress continues, then it is undoubtedly true that most of the beneficiaries of a particular vaccine, etc., are simply just statistics. Unfortunately, there are those who are quick to focus attention on the victim, not as a show of concern but as an argument against progress, while implicitly or explicitly denying its benefits.

It would be nice to have progress without cost, but that is simply not possible. It is possible, however, to reduce the costs of progress. Those most likely to denigrate science and scientists are often the same ones who demand the impossible: Science as a human endeavor free of all possible error. Knowledge is cumulative and from our past mistakes we can learn to avoid similar mistakes and overall reduce the error in the process. If critics would focus on improving the process to reduce error instead of issuing blanket condemnations, then there would be fewer victims. In any process we will always need to be careful in evaluating human costs and benefits to make sure that procedures to reduce error do not go beyond a point where they delay or otherwise significantly reduce a greater benefit. The process must be just in that those who most likely benefit are those who are at risk of the harm unless they, for one reason or another, freely and with full knowledge of risks involved, volunteer to be in harm's way. Any other skewing of the process, where some are regularly at risk while others are more likely to benefit, raises questions of justice which need to be addressed. Certainly this is an aspect of the process as currently being carried out where criticism should be leveled and improvement sought. For all concerned and involved, there must be as complete a transparency as humanly possible. No matter how much tested, regulated, and studied, any pharmaceutical, pesticide, industrial chemical, etc., may be, given the limits of human knowledge, there will always be an unknown and the possibility of unintended harm.

Whatever proof of safety is given, there will always be the cry for one more test, or the claim against modern technology that not enough testing has been carried out. What the critics really want is continuation of testing until one has findings compatible with their preconceptions, and then they will be more than willing to act upon it and force the rest

of us to do so as well. This is another standard charge against genetically modified foods by those who have no idea how much experience
and testing has already occurred.

Given the importance of nutrition to infants and children, one would
not expect the results that Fogel finds if our modern food supply was
being contaminated by pesticides and other toxic chemicals that were
threatening our health and well being. Yet the thrust of much of our
regulatory legislation has been on the basis of protecting children in
spite of findings such as the following:

> During the past 50 years of regulating thousands of substances, there is
> no known case of toxicity in children from the ingestion of food addi-
> tives or pesticides that were used in conformity with established toler-
> ances (Scheuplein 2000, 275).

Not that there have not been any problems from the use of these
chemicals but they are the result of "accidental exposures, intentional
abuse, illegal use, and exposure to applicators or to farm workers." It
was this type of misuse of these "chemicals" which explains the "entire
inventory of cases of human toxicity to pesticides" (Scheuplein 2000,
275). It is just possible that "chemicals" are not killing us.

In addition to higher levels of nutrition and cleaner, safer food, modern consumers now have an incredible array of foodstuffs from around
the world as well as an opportunity to savor, with some frequency,
cuisines from cultures whose culinary delights were unknown to their
parents or grandparents. In an article appropriately titled, "Mean
Cuisine," Greg Critser asks the question "Why, in a time of unprecedented abundance for everyone—vine-ripened Mexican tomatoes for
$1 a pound! World-class reds and whites from Montepulciano
d'Abruzzo for $5 a bottle! An international glut of inexpensive extra
virgin olive oils and cheeses and nuts and fruits at Trader Joe's and
Price Club!—why oh why are the chefs of America so dour, so chary—
so very very very bummed out?" "Why the big change" Critser asks?
"Ten years ago, a pint of cold-pressed, extra-virgin Italian olive oil
would set you back about $20. It was scarce, and so it was the chef's
preference. Today one can buy a gallon for the same price. Today, of
course, imported oil is not the chef's choice" (Critser 2001b). The
answer is abundance, and abundance is a threat to the values of snobbery of the critics of modernity.

The snobfest itself flows from what the great historian Richard Hofstadter called "status anxiety," the sinking feeling, often felt after, say, actually speaking to the maid or the gardener, that the world is changing, expanding, and in the process making one smaller, less important (Critser 2001b).

Critser adds that the "culprit is globalization." The foods, particularly those that were once imported at a price beyond the reach of ordinary citizens, have now become common and relatively cheap in supermarkets across the land. Globalization has been the mechanism by which the increasing global food production provides greater diversity of available foodstuffs and therefore greater choice, but it also deprives the elitist of that sense of exclusivity for the items which they consume. Technology has made for improvement and greater availability of high quality items such as fine wines (Feiring 2001). In a world of increasing free trade and technological advancement, the food snobs seek to pursue an antitrade ("buy locally"), antitechnology agenda in order to preserve their status and self-esteem, even if it is at the expense of continuing the increase in food production to meet a growing world population and make the technologies of accessibility and "technologies of abundance" available to those who have not had the opportunity to benefit from them as fully as others have. Rules that make items of consumption more expensive, restrict access to them to those who can afford them, and thereby make them more prestigious.

Modernity, rather than giving us the global homogenization that many feared and continue to fear, has given us unprecedented freedom of choice in the foods we eat, the music to which we listen, along with about every other aspect of life, and it has given us the health and longevity to appreciate them. Modernity has even given its critics the affluence to engage in their expensive fetishes of consumption and snobbery which is their privilege. None of us begrudges them this right and privilege. But when they seek to impose their antimodernist agenda on the rest of us, then it is essential that we stop and take stock of what we have achieved, how far we have come, the potential of where we can go, and the need to defend and preserve that which those who went before have won for us and what we owe to those who follow in our footsteps. The continuing promise of modernity is not only about what we have already achieved but about the victories yet to be won. The promise of modernity is one of breaking down barriers of all kinds: To trade, to the exchange of ideas, to the appreciation of the cultural achievements of others, and even to the recognition of there being "others." Here is an even more important underlying sense of a shared humanity.

The edifice of technology is integral to, and inseparable from, the larger edifice of human knowledge and culture. Technologies of accessibility and opportunity are a product of advancing knowledge and its widest possible dissemination. To some of us, there is a faith that has broader, beneficial implications for the entirety of the human endeavor. As John Dewey maintained, "The formation of a cultivated and effectively operative good judgment or taste with respect to what is aesthetically admirable, intellectually acceptable and morally approvable is the supreme task set to human beings by the incidents of experience." For it is in the unity of experience and the accumulation of knowledge that we manifest the true character of ourselves. "There is nothing in which a person so completely reveals himself as in the things which he judges enjoyable and desirable. Such judgements are the sole alternative to the domination of belief by impulse, chance, blind habit and self-interest." Fundamentally, then, our "relatively immediate judgements ... do not precede reflective inquiry but are the funded product of much thoughtful experience." "Expertness of taste is at once the result and the reward of constant exercise of thinking. Instead of there being no disputing about tastes, they are the one thing worth disputing about, if by 'dispute' is signified discussion involving reflective inquiry. Taste, if we use the word in its best sense, is the outcome of experience brought cumulatively to bear on the intelligent appreciation of the real worth of likings and enjoyment" (Dewey 1929, 262).

The technologies of accessibility are "technologies of opportunity," as their very accessibility gives new opportunities for personal growth and development to an ever larger portion of the population. If we can overcome snobbery and elitism, then affluence and the moving forward of the frontiers of knowledge can facilitate technologies of accessibility and opportunity continuing to be "technologies of excellence" as new niches for creativity arise and more human potential and talent are unleashed—the poet's "mute, inglorious Miltons" will be given a voice. Mass culture need not be about lowering us all to the proverbial "least common denominator" but continually raising the denominator for all. Going beyond the democratic glorification of the "common man," the potential is to create conditions so that which is now uncommon becomes common.

The true promise of modernity and the technologies of accessibility is inclusivity. Consequently, to some of us, modernity's greatest attainment, which is still a distant dream, is when *all* who wish to partake of its benefits, have the opportunity to do so. However distant, it is a promise that can be kept, it is a promise towards which we must be striving, and it is a promise that *must* be kept. The true promise of modernity is a better life for all who want it.

References

ABC. 1995. "Prime Time Live," American Broadcasting Company, 2 August.

Abreu, Robin. 1994. Keen to be Green. *India Today* 29(17):82–83, 15 September.

Ackerman, Diane. 1990. *A Natural History of the Senses.* New York: Random House.

Adams, Jonathan S., and Thomas O. McShane. 1996. *The Myth of Wild Africa: Conservation Without Illusion.* Berkeley: University of California Press.

Adams, Perry G. 1983. *Travel Literature and the Evolution of the Novel.* Lexington: The University of Kentucky Press.

Adhikari, Gautarn. 1988. Yes, Mr. Buckley, There's an Indian Literature. *New York Times*, 10 June.

Adorno, Theodor. 1973. *Philosophy of Modern Music.* Translated by Anne G. Mitchell and Wesley V. Blomster. New York: The Seabury Press, A Continuum Book.

AFP. 2001. New Release of "The Lion Sleeps Tonight." *Agence France-Presse, Daily Mail & Guardian* online, 3 August.

Agarwal, Anil, and Sunita Narain. 1996. Spices of Life. *Asia Times* 266:ll, 19 December. (Reprinted from *New Scientist*).

Agbiotech Bulletin. 2001. High Anxiety and Biotechnology: Who's Buying, Who's Not, and Why? *Agbiotech Bulletin* 9(5), June.

Alexie, Sherman. 1995. *Reservation Blues.* New York: The Atlantic Monthly Press.

Allaby, Michael, and Jim Lovelock. 1980. Wood Stoves: The Trendy Pollutant. *New Scientist* 88(1228):420–422, 13 November.

Allan, N. 1988. Highways to the Sky: The Impact of Tourism on South Asian Mountain Culture. *Tourism Recreation Research* 13:11–16.

Allen, Jim. 1997. The Impact of Pleistocene Hunters and Gatherers on the Ecosystems of Australia and Melanesia: In Tune With Nature? In *Historical Ecology in the Pacific Islands: Prehistoric Environmental and Landscape Change*, edited by Patrick V. Kirch and Terry L. Hunt. New Haven: Yale University Press.

Allison, Marvin J. 1984. Paleopathology in Peruvian and Chilean Populations. In *Paleopathology at the Origins of Agriculture*, edited by Mark Cohen and George J. Armelagos. New York: Academic Press.

Ames, Bruce N. 1983. Dietary Carcinogens and Anticarcinogens: Oxygen Radicals and Degenerative Diseases. *Science* 221(4617):1256–1264, 23 September.

Ameyibor, Edward. 1997. Ghana's Game Rangers Call a Truce With Villagers. *Electronic Mail & Guardian* online, 14 October.

ANAPQUI (National Association of Quinoa Producers, Bolivia). 1997. Bolivian Farmers Demand Researchers Drop Patent on Andean Food Crop. Press Release on the Internet, 20 June.

Anderson, Atholl. 1997. Prehistoric Polynesian Impact on the New Zealand Environment: Te Whenua Hou. In *Historical Ecology in the Pacific Islands: Prehistoric Environmental and Landscape Change*, edited by Patrick V. Kirch and Terry L. Hunt. New Haven: Yale University Press.

Anderson, John Ward. 2001. Roll Over, Khomeini! Iran Cultivates A Local Rock Scene, Within Limits: Iranian Pop Star Shadmehr Aghili Has Never Been Allowed by the Government to Give a Live Concert. *Washington Post*, August 23.

Andrews, Malcolm. 1989. *The Search for the Picturesque: Landscape Aesthetics and Tourism in Britain, 1760–1800*. Stanford, CA: Stanford University Press.

AP. 1990. Earth-friendly Snack Appeals to Consumers. *Houston Post*, 19 December.

———. 1992. "Coca Colonizing" the World: Good? Bad? Inevitable? *Sunday Observer* (Jakarta, Indonesia), 22 March.

———. 1997. South Pacific Forum to Consider Dose of Strong Economic Medicine. *Nando.net* online, 16 September.

———. 1998. Dog Food Recalled After Pets Sicken, Die. *Houston Chronicle*, 4 November.

Aperture. 1990. Socially Responsible Collecting (an Ad for *Mother Jones'* Fine Print Program). *Aperture* (119):80, Summer.

Ardrey, Robert. 1963. *African Genesis: A Personal Investigation into Animal Origins and Nature of Man*. New York: Delta Publishing Company.

Armstrong, Sue. 1991. The People Who Want Their Parks Back. *New Scientist* 131(1776):54–55, 6 July.

———. 1996. A Landless People of the Land. *Electronic Mail & Guardian* online, 11 July.

Armstrong, Virginia Irving. 1971. *I Have Spoken: American History Through the Voices of the Indians*. Chicago: The Swallow Press.

Arnold, Steven F., Diane M. Klotz, Bridgette M. Collins, Peter M. Vornier, Louis J. Guilette Jr., and John A. McLachlan. 1996. Synergistic Activation of Estrogen Receptor with Combinations of Environmental Chemicals. *Science* 272(5267):1489–1492, 7 June.

ART (African Resources Trust). 1997. CITES Meeting in Harare: Implications for the ACP Countries. *The ACP-EU Courier* (165):9–10, September–October.

Asiaweek. 1995. Born in the PRC: China's New Voices of Rock—Mandarin Rock Shakes Up Asia's Music Scene. *Asiaweek* 21(20)38–43, 19 May.

————. 1997. McChina: The Americanization of China. *Asiaweek* (26), 4 July.

Athens, J. Stephen, and Jerome V. Ward. 1993. Environmental Change and Prehistoric Polynesian Settlement in Hawaii. *Asian Perspectives: The Journal of Archaeology for Asia and the Pacific* 32(2):205–223, Fall.

Ayres, Clarence E. 1961. *Toward a Reasonable Society: The Values of Industrial Civilization.* Austin, Texas: University of Texas Press.

B&J (Ben & Jerry). 1999. Ben & Jerry's Thoughts on Dioxin. *Ben & Jerry's* online, November.

Baeyer, Hans Christian von. 2000. The Lotus Effect. *Sciences* 40(1):12–15, January/February.

Bagley, Clarence B. 1931. Chief Seattle and Angeline. *Washington Historical Quarterly* 22(4):243–272, October.

Bahn, Paul, and John Flenley. 1992. *Easter Island, Earth Island.* New York: Thames and Hudson.

Bahro, Rudolph. 1986. Building the Green Movement. Translated by Mary Tyler. London: G.M.P.

Bailey, Ronald. 2001, Dematerializing the Economy, *Reason*, September.

Bakker-Cole, Mary. 1995. Zimbabwe's Elephants Up for Sale. *New Scientist* 146(1973):6, April.

Baldwin, J. 1985. Equipment for the New Age. *Outside* 10(1):57–66, January.

Balter, Michael. 2001. Paleoanthropology: What—or Who—Did in the Neanderthals?, *Science* 293(5537):1980–1981, 14 September.

Barden, Carol Isaak. 1991. Visit to the Dark Continent is Spiritual Impression. *Houston Post*, 13 January.

Barrett, Mary Ellen. 1990. Peddling the Planet. *USA Weekend*:4–5, 14–16 December.

Barsh, Russel. 1990. Cattle on the Great Plains. In *The Struggle for Land: Indigenous Insight and Industrial Empire in the Semiarid World*, edited by Paul A. Olson. Lincoln, NB: University of Nebraska Press.

Bates, Marston. 1967. *Gluttons and Libertines: Human Problems of Being Natural.* New York: Vantage Books.

BBC. 1998. Thailand to Alter Gas Pipeline Route. *BBC World Service,* online 6 January.

————. 1999a. A Taste for Meat. *BBC World Service,* online, 15 January.

————. 1999b. S. Africa Returns Land to Indigenous Tribe. *BBC World Service,* online 21 March.

————. 1999c. Vegan Condoms Launched. *BBC World Service* online, 16 April.

————. 2001. Intelligence Linked to Birthweight: Nutrition During Pregnancy May Affect IQ. *BBC World Service* online, 9 August.

Beament, Sir James. 1997. *The Violin Explained: Components, Mechanism and Sound.* Oxford: Clarendon Press.

Beattie, Alan. 2001. Campaigners Offer Moral Integrity for Influence. *Financial Times,* 17 July.

Beinart, William. 1989a. Introduction: The Politics of Colonial Conservation. *Journal of Southern African Studies* 15(2):143–162, January. (Special Issue: *The Politics of Conservation in Southern Africa*).

————, ed. 1989b. *Journal of Southern African Studies* 15(2), January. (Special Issue: *The Politics of Conservation in Southern Africa*).

Bell, R.H.V. 1987. Conservation with a Human Face: Conflict and Reconciliation in African Land Use Planning. In *Conservation in Africa: People, Policies and Practice*, edited by David Anderson and Richard Gove, Cambridge: Cambridge University Press.

Bell, Susan. 1997. French Conservationists Contend that Chanel Perfume Threatens Rain Forest. *Times of London News Service, Nando.net* online, 2 July.

Bernardes, Ernesto. 1997. Article in *Veja*, 15 January. Reprinted *World Press Review* 34(4):32–33, April.

Berreman, Gerald D. 1991. The Incredible 'Tasaday': Deconstructing the Myth of a 'Stone-Age' People. *Cultural Survival Quarterly* 15(1).

————. 1992. The Tasaday: Stone Age Survivors or Space Age Fakes? In *The Tasaday Controversy: Assessing the Evidence*, edited by Thomas N. Headland. Washington: American Anthropological Association, Scholarly Series No. 28.

Berry, Patricia. 1982. *Echo's Subtle Body: Contributions to an Archetypical Psychology*. Dallas, TX: Spring Publications.

Bester, Roy, and Barbara Buntman. 1999. Bushman(ia) and Photographic Intervention. *African Arts* 32(4):50–59, Winter.

Binns, Archie. 1944. *Northwest Gateway: The Story of the Port of Seattle*. Garden City, NY: Doubleday, Doran and Company, Inc.

Bishop, Peter. 1989. *The Myths of Shangri-La Tibet: Travel Writing and the Western Creation of Sacred Landscape*. London: The Atherton Press.

Bittman, Mark. 1994. Eating Well: A Little Cooking Goes a Long Way to Make the Most of Vegetable Nutrients. *New York Times*, 31 August.

Black Elk. 1971. The Sacred Pipe: Black Elk's Account of the Seven Rites of the Oglala Sioux. Recorded and edited by Joseph Epes Brown. Baltimore, Md.: Penguin Books. Originally published in 1953.

Black, Johnny. 1993. Roll Over Chuck Berry. In *Q*, Issue No. 78, March.

Blainey, Geoffrey. 1976. *Triumph of the Nomads: A History of Ancient Australia*. South Melbourne, Victoria, Australia: Sun Books.

Blesh, Rudi, and Harriet Janis. 1971. *They All Played Ragtime*. New York: Oak Publishers.

Bogan, S., and F. Williams. 1991. WWF Bankrolled Rhino Mercenaries. *Independent on Sunday* (London), 17 November.

Bond, Michael. 2001. Off Centre: Had the Wild West Already Been Tamed?: Claims That The Landscape Had Been Extensively Altered By Humans Long Before Europeans Arrived Is Disturbing The American Dream, *Financial Times*, 30 June.

Bonner, Raymond. 1993a. *At the Hand of Man: Peril and Hope for Africa's Wildlife*. New York: Knopf.

———. 1993b. Crying Wolf over Elephants. *New York Times Magazine*: 16–19, 30, 40–53, 7 February.

———. 1994a. Western Conservation Groups and the Ivory Ban Wagon. In *Elephants and Whales: Resources for Whom?*, edited by Milton M.R. Freeman and Urs P. Kreuter. Basel, Switzerland: Gordon and Breach Science Publishers.

———. 1994b. Compassion Wasn't Enough in Rwanda. *New York Times*, Section 4, Week in Review: 3, 18 December.

Bose, Kunal. 1994. India Opts for Organic Cotton. *Financial Times* (London), 16 September.

Boswell, John. 1988. *The Kindness of Strangers: The Abandonment of Children in Western Europe from Late Antiquity to the Renaissance*. New York: Pantheon Books.

Bower, Bruce. 1988. Murder in Good Company: Cooperationary Comradery and a Dizzying Homicide Rate Distinguish a Small New Guinea Society. *Science News* 133(6):90–91, February 6.

Bradley, Robert. 1997. *Problems and Controversies of Renewable Energy*. Draft Manuscript, 10 February.

Bramwell, Anna. 1989. *Ecology in the 20th Century: A History*. New Haven, CT: Yale University Press.

Brent, Michel. 1996. A View Inside the Illicit Trade in African Antiquities. In *Plundering Africa's Past*, edited by Peter Schmitt and Roderick J. McIntosh. Bloomington: Indiana University Press.

Brightman, Robert A. 1987. Conservation and Resource Depletion: The Case of the Boreal Forest Algonquins. In *The Question of the Commons: The Culture and Ecology of Communal Resources,* edited by Bonnie M. McCay and James M. Acheson. Tucson: University of Arizona Press.

Brockington, Daniel, and Katherine Homewood. 1996. Wildlife, Pastoralists and Science. In *The Lie of the Land: Challenging Received Wisdom on the African Environment,* by Melisa Leach and Robin Mearns. Oxford: The International African Institute in Association with James Currey and Heinemann.

Brooks, David. 1999. How You Can Buy Old-Fashioned Virtues. *New Yorker* 74(43):36–41, 25 January.

Brown, Patricia Leigh. 1990. A 'Healthy House' That's High in Style but Low in Chemicals. *New York Times,* 19 April.

Brown, Paul. 2000. Traders Say Profit Can be Motive for Preservation. *Electronic Mail & Guardian* online, 7 April.

Brown, Richard P. C. 1998. Do Migrants' Remittances Decline Over Time?: Evidence from Tongans and Western Samoans in Australia. *The Contemporary Pacific: A Journal of Island Affairs* 10(1):107–151, Spring.

Bruner, Edward M. 1996. Tourism in the Balinese Border Zone. In *Displacement, Diaspora, and Geographies of Identity*, edited by Smadar Lavie and Ted Swedenburg. Durham, NC: Duke University Press.

Brunton, Ron. 1995. Indigenous People Live in Harmony with Nature and Have Much to Teach Us About Environmental Stewardship. In *Tall Green Tales*, edited by Jeff Bennett. Western Australia: Institute of Public Affairs.

Brush, Stephen B. 1993. Indigenous Knowledge of Biological Resources and Intellectual Property Rights: The Role of Anthropology. *American Anthropologist* 95(3):653–671, September.

Bryant, Paul. 1989. "Nature Writing and the American Dream." In *The Frontier Experience and the American Dream: Essays on American Literature*, edited by David Mogen, Mark Busby, and Paul Bryant. College Station, TX: Texas A&M University Press.

Budiansky, Stephen. 1995a. Chaos in Eden, *New Scientist* 148(1999):33–36, 14 October.

———. 1995b. *Nature's Keepers: The New Science of Nature Management.* New York: Free Press.

Burnett, Andrea B., and Larry R. Beuchat. 2001. Comparison of Sample Preparation Methods for Recovering Salmonella from Raw Fruits, Vegetables, and Herbs. *Journal of Food Protection* 64(10):1459–1465, October.

Buzzworm. 1990. Advertisement, *Buzzworm: The Environment Journal* 2(4):9, 17, 21, July/August.

Caldwell, John, Pat Caldwell, and Bruce Caldwell. 1987. Anthropology and Demography: The Mutual Reinforcement of Speculation and Research. *Current Anthropology* 28(1):25–43, February.

Callicott, J. Baird. 1989. American Indian Land Wisdom. *Journal of American Forestry* 35–42, January. Reprinted in *The Struggle for Land: Indigenous Insight and Industrial Empire in the Semiarid World*, edited by Paul A. Olson. Lincoln, NB: University of Nebraska Press.

———. 1991. The Wilderness Idea Revisited: The Sustainable Development Alternative. *The Environmental Professional* 13(2):235–247.

Carey, Susan. 2001. The Virtuous Vacation? *Wall Street Journal*, 27 July.

Carlson, Peter. 2000. Conspicuous Simplicity for Beautiful People: Real Simple Only Faintly a Throwback to Thoreau. *Washington Post*, 11 April.

Carruthers, Jane. 1989a. *Creating a National Park, 1910 to 1926*. Johannesburg, South Africa: University of the Witwatersrand, African Studies Institute.

———. 1989b. Creating a National Park, 1910–1926. *Journal of Southern African Studies* 15(2):188–216, January. (Special Issue: *The Politics of Conservation in Southern Africa*).

———. 1994. Dissecting the Myth: Paul Kruger and the Kruger National Park. *Journal of Southern African Studies* 20(2):263–283, January.

———. 1995a. *The Kruger National Park: A Social and Political History*. Pietermaritzburg, South Africa: University of Natal Press.

———. 1995b. *Game Protection in the Transvaal 1846–1926*. Archives Yearbook for South African History, Pretoria, South Africa: The Government Printer.

———. 1999. Protected for the People or Against the People? National Parks and Game Reserves in the Transvaal and Natal. Perth, Western Australia: African Studies Centre of Western Australia, African Seminar Program, 5 February.

Cartmill, Matt. 1993. *A View to a Death in the Morning: Hunting and Nature Through History*. Cambridge, MA: Harvard University Press.

Cassidy, John. 1999. No Satisfaction: The Trials of the Shopping Nation. *New Yorker* 74(43):88–92, 25 January.

Castles, Ian. 2000. Reporting on Human Development: Lies, Damned Lies and Statistics, Perspectives On Global Economic Progress And Human Development. Academy of Social Sciences in Australia, Occasional Paper Series 1/2000.

CBC. 2001. Dead Bodies Could Enrich Soil Faster. Canadian Broadcasting Corporation, online, 1 June.

Chadwick, Douglas H. 1992. *The Fate of the Elephant*. San Francisco: Sierra Club Books.

Chagnon, Napoleon A. 1988. Life Histories, Blood Revenge, and Warfare in a Tribal Population. *Science* 239(4848), 26 February.

Charles, Daniel. 2001a. Agbiotech: Seeds of Discontent, Plant Breeders, Developing Nations, and Agricultural Firms Battle for Control of the World's Stock of Crop Diversity. *Science* 294(5543):772–775, 26 October.

———. 2001b. Seed Treaty Signed: U.S., Japan Abstain. *Science* 294(5545):1263-1264, 9 November.

Chase, Alston. 1986. *Playing God in Yellowstone: The Destruction of America's First National Park*. Boston: Atlantic Monthly Press.

Churchill, Steven E. 2001. Commentary: Hand Morphology, Manipulation, and Tool Use in Neanderthals and Early Modern Humans of the Near East. Proceedings of the National Academy of Sciences of the United States of America 98(6):2953–2955, 13 March.

Clarke, William C. 1971. *Place and People: An Ecology of a New Guinean Community.* Canberra: Australia University Press.

———. 1990. Learning from the Past: Traditional Knowledge and Sustainable Development. *The Contemporary Pacific: A Journal of Island Affairs* 2(2):233–253, Fall.

Clay, Jason. 1992. Why Rainforest Crunch. *Cultural Survival Quarterly* 16(2):31–37, Spring.

Clifton, James A., ed. 1990. *The Invented Indian: Cultural Fictions and Government Policies.* New Brunswick: Transaction Publishers.

Closmann, Charles E. 1997. *Nature Protection in Nazi Germany: The Search for National Identity in Nature.* Master's thesis, University of Houston, n.d.

Coe, Michael D. 1993. "From Huaquero to Connoisseur: The Early Market in Pre-Columbian Art." In *Collecting the Pre-Columbian Past,* a symposium at Dumbarton Oaks, 6th and 7th October 1990, edited by Elizabeth Hill Boone. Washington: Dumbarton Oaks Research Library and Collection.

Cohen, Mark N. 1977. *The Food Crisis in Prehistory: Overpopulation and the Origins of Agriculture.* New Haven, CT: Yale University Press.

———. 1987. The Significance of Long-term Changes in Human Diet and Food Economy. In *Food and Evolution: Toward a Theory of Human Food Habits,* edited by Marvin Harris and Eric B. Ross. Philadelphia: Temple University Press.

Cohen, Patricia. 1998. Sociologists with a Gig Off the Beat. *New York Times,* 29 August.

Concar, David, and Mary Cole. 1992. Commerce and the Ivory Tower. *New Scientist* 133(1810):29–33, 20 February.

Cook, Scott. 1974. Structural Substantivism, review of *Stone Age Economics,* by Marshall Sahlins. *Comparative Studies in Society and History: An International Quarterly* 16(4):355–379, September.

Coontz, Stephanie. 1992. *The Way We Never Were: American Families and the Nostalgia Trap.* New York: Basic Books.

Corbain, Alain. 1994. *The Lure of the Sea: The Discovery of Seaside in the Western World, 1750–1840.* Translated by Jocelyn Phelps. Oxford: Polity.

Cowley, Geoffrey. 1989. The Death of an Illusion: Bill McKibben's Mournful Reflection on the Environment is Based on Romantic Fantasy. *Newsweek* 114(17):83, 23 October.

Craig, Jacqu. 2001. Conservation Versus Development. *Spiked-Online,* http://www.spiked-online.com, 2 August.

Cray, Dan. 1998. Navajo vs. Navajo: A Battle Over Whether to Preserve Natural Resources or Develop Them. *Time* 152(4), 27 July.

Critser, Greg. 2001a. Forget Organic: Just Eat Those Veggies, *Los Angeles Times*, 20 May.

————. 2001b. Mean Cuisine: Gone Is The Joy of Cooking. Today's Celebrity Chefs Are Serving Up a Menu of Global Doom and Politically Twisted Snobbery. *Washington Monthly*, July/August.

Croll, Elisabeth J., and David J. Parkin. 1992. Anthropology, the Environment and Development. In *Bush Base: Forest Farm: Culture, Environment, and Development,* edited by Elizabeth J. Croll and David J. Parkin. London & New York: Routledge.

Cronon, William. 1995a. The Trouble with Wilderness: Wilderness is No More 'Natural' than Nature Is—It's a Reflection of Our Own Longings, a Profoundly Human Creation. *New York Times Magazine*, August 13:42–43.

————. 1995b. *Uncommon Ground: Toward Reinventing Nature.* New York: W.W. Norton.

————. 1996. The Trouble with Wilderness, or Getting Back to the Wrong Nature. *Environmental History* 1(1):7–25, January.

Crossette, Barbara. 1989. India's Homespun Chic, No Imports. *New York Times*, 19 December.

CSE (Centre for Science and Environment). 1986. *The State of India's Environment—1984–85—The Second Citizen's Report.* Delhi, India: The Centre for Science and Environment.

Currey, B. 1980. Famines in the Pacific: Losing the Chances for Change. *Geojournal* 4(5):447–466.

Daley, Suzanne. 1997a. Muzarabi Journal: Where Elephants Pay Their Way. *New York Times,* 12 April.

————. 1997b. Ban on Sale of Ivory is Eased to Help 3 African Nations: Sale from Stockpiles of Ivory Will Be First in More Than Seven Years. *New York Times,* 20 June.

Dalton, Rex. 2000. Cereal Gene Bank Accepts Need for Patents. *Nature* 404(6778):534, 6 April.

Davies, Caitlin. 1998. Botswana's Basarwa Remain Poor and Marginalised. *MISAnet/Africa Information Afrique* online, 21 September.

Davies, Frank. 1999. Amazon Alliance Protests Patent on Sacred Plant *Miami Herald*, 31 March.

Davis, Francis. 1995. *The History of the Blues: The Roots, the Music, the People from Charley Patton to Robert Cray.* New York: Hyperion.

Debroy, Bibek. 2001. Listen to Inventors, Not Activists. *Business Day* (Johannesburg), 8 November.

DeGregori, Thomas R. 1985a. *A Theory of Technology: Continuity and Change in Human Development.* Ames, IA: Iowa State University Press.

————. 1985b. Technological Limits to Forecasts of Doom: Science, Technology, and the Sustainable Economy. *Technovation* 3:209–220.

————. 1986. Redefining the Agenda: Resource Creation and Open Ended Development. *Forum for Applied Research and Public Policy* 1(2):40–44, Summer 1986.

————. 1987a. Population, Technology, Cognition, and Resource Creation: Humanizing the Environment for Habitat and Higher Achievement. Paper presented at a Conference of the European Population Union in Jyvaskyla, Finland, June.

————. 1987b. Resources Are Not; They Become. *Journal of Economic Issues* 20(2):463–470, June.

————. 1989. Goodbye to Nature: A Compendium of New Age Nonsense! *Houston Chronicle*, 12 November.

————. 1995. Natural Resources and Sustainable Development. In *Economic Policy and the Environment,* edited by Mark Griffith and Bishnodat Persaud. Mona, Jamaica: University of the West Indies, Centre for Environment and Development (UNICED).

————. 1996. Technology Transfer, Economic Development, and the Perpetuation of Poverty: Resource Creation Versus Frugality. In *The Institutional Economics of the International Economy,* edited by John Adams and Anthony Scaperlanda. Boston: Kluwer Academic Publishers.

————. 1998. Back to the Future?: A Review Article. *Journal of Economic Issues* 32(4), December.

————. 2001. *Agriculture and Modern Technology: A Defense.* Ames, IA: Iowa State University Press.

DeLong, J. Bradford. 1991–2000. The Economic History of the Twentieth Century: Slouching Towards Utopia. online www.j-bradford-delong. net.

Deloria, Vine, Jr. 1979. Introduction. *Black Elk Speaks: Being the Life Story of a Holy Man of the Oglala Sioux,* by Black Elk as told through John G. Neihardt. Lincoln, NE: University of Nebraska Press.

de Mille, Richard. 1990. "Validity is Not Authenticity: Distinguishing Two Components of Truth." In *The Invented Indian: Cultural Fictions and Government Policies*, edited by James A. Clifton, pp. 227–253. New Brunswick: Transaction Publishers.

Denevan, William M. 1992. The Pristine Myth: The Landscape of the Americas in 1492. *Annals of the Association of American Geographers* 82(3):369–385, September, Karl W. Butzer, Guest Editor.

Dennett, Glen, and John Conell. 1988. Acculturalism and Health in the Highlands of Papua New Guinea: Dissent and Diversity Needs and Development. *Current Anthropology* 29(2):273–282, April.

Denny, Charlotte. 2001. Nile Power Row Splits Uganda: Africans Want Environmentalists Out of Their Backyard So Dam Project Can Light Their Evenings. *Guardian* (London), 15 August.

DePalma, Anthony. 2000. Texcoco Journal: The 'Slippery Slope' of Patenting Farmers' Crops, *New York Times*, 24 May.

Dermansky, Ann. 1991. What's in Store: It's Easier Being Green— Eco-Boutiques from Soho to Market Street. *ElleDecor* 2(4):64, May.

de Sola Pool, Ithiel. 1979. Direct Broadcast Satellites and the Integrity of National Cultures in National Sovereignty and International Communication, edited by K. Nordenstreng and H.L. Schiller. New Jersey: Ablex, 1979.

Dewey, John. 1929. *The Quest for Certainty: A Study of the Relation of Knowledge and Action.* Reprint, New York: Capricorn Books, G.P. Putnam & Sons, 1980.

————. 1934. *Art as Experience.* New York: Minton, Balch & Co.

————. 1958. *Experience and Nature.* New York: Dover Publications.

Diamond, Jared M. 1994. Ecological Collapses of Ancient Civilizations: The Golden Age That Never Was. *The Bulletin of the American Academy of Arts and Sciences* 37–59, February.

————. 1997. *Guns, Germs and Steel: The Fates of Human Societies.* New York: W.W. Norton.

————. 2000. Archaeology: Talk of Cannibalism. *Nature* 407(6800):25–26, 7 September.

Dilworth, Leah. 1996. *Imagining Indians in the Southwest: Persistent Visions of a Primitive Past.* Washington: Smithsonian Institution Press.

Dominick, Raymond H. III. 1992. *The Environmental Movement in Germany: Prophets and Pioneers, 1871–1971.* Bloomington, IN: Indiana University Press.

Dorsey, Michael K. 1998. Toward an Idea of International Environmental Justice. In *World Resources 1998–99 A Guide to the Global Environment: People and the Environment: Environmental Change and Human Health.* New York: Oxford University Press for the World Resources Institute.

Dove, Michael. 1994. *Marketing the Rainforest: Green Panacea or Red Herring?* Asia-Pacific Issues: Analysis from the East-West Center, No. 13. Honolulu: East-West Center.

Drury, William H. 1998. *Chance and Change: Ecology for Conservationists.* Berkeley: University of California Press.

Dubos, Rene, and Jean Paul Escande. 1980. *Quest: Reflections on Medicine, Science and Humanity.* Translated by Patricia Ramum. New York: Harcourt, Brace, Jovanovich.

Du Noyer, Paul. 1993. You Hum It, Son. *Q* 78:49–51, March.

Durbin, Paul T. ed. 1984. *A Guide to the Culture of Science, Technology and Medicine.* New York: Free Press.

Durham, Michael, and Jan Rocha. 1996. Amazon Chief Sues Body Shop: Anita Roddick Exploited Me, Claims Village Elder. *Observer* (London), 3 March.

Earth Care Paper Inc. 1990. *Recycled Paper Catalog*. Madison: Earth Care Paper Inc., Fall/Winter.

Easterbrook, Gregg. 1989. It's Not Nice to Fool Mother Nature. *Washington Monthly* 21(9):51–54, October.

Economist. 1997a. Surprise in the Woods. *Economist* 344(8026):31–32, 19 July.

———. 1997b. The Rhinos' Return: Shooting the Shooters. *Economist* 344(8035):96, 20 September.

———. 2001a. Special Report: Human Rights, Righting Wrongs. *Economist* 360(8235):19–21, 18 August.

———. 2001b. Lexington: Leon Kass, Philospher-politician. *Economist* 360(8235):41, 18 August.

Eder, James F. 1994. State-Sponsored Participatory Development and Tribal Filipino Ethnic Identity. *Social Analysis* 14:28–38.

Edwards, Rob. 1997. Beware Green Imperialists: In Their Zeal to Preserve Endangered Species for the Good of the Global Environment, Conservationists and Scientists are Ignoring the Needs of Indigenous Peoples. *New Scientist* 14–15, May 31.

Egan, Timothy. 1998. An Indian Without Reservations. *New York Times Magazine*, 16–19, 18 January.

Ehrenreich, Barbara. 1999. Men Hate War Too. *Foreign Affairs* 78(1):118–122, January/February.

Ehrlich, Cyril. 1988. History of the Piano: 1860–1915. In *The New Grove Dictionary of Musical Instruments: Piano*, edited by Stanley Sadie. New York: W.W. Norton and Company.

———. 1990. *The Piano: A History*. New York: Oxford University Press.

Ehrlich, Paul R., and Anne H. Ehrlich. 1981. *Extinctions: The Causes and Consequences of the Disappearance of Species*. New York: Random House.

———. 1990. *The Population Explosion*. New York: Simon and Schuster.

Elgin, Duane. 1981. *Voluntary Simplicity: Toward a Way of Life That is Outwardly Simple, Inwardly Rich*. New York: William Morrow and Co., Inc.

Ellis, Stephen. 1994. Of Elephants and Men: Politics and Nature Conservation in South Africa. *Journal of Southern African Studies* 20(1):53–69, March.

Emsley, John. 2001. Good News is No News. *Nature* 413(6852):113, 13 September.

Eng, Peter. 1998. A New Kind of Cyberwar: In Burma, Thailand, Indonesia: Bloodless Conflict. *Columbia Journalism Review* 37(3):20–21, September/October.

Entine, Jon. 1995. Rainforest Chic, *The Toronto Globe and Mail Report on Business Magazine*, 40–52, October.

———. 1996a. The Messy Reality of Socially Responsible Marketing. Better Worlds Web Site. Reprinted from *Utne Reader*, 1995 and *At Work*, May/June 1995.

————. 1996b. Let Them Eat Brazil Nuts. *Dollars and Sense*, 204:30–35, March/April.

Epstein, Jack. 1993. Brazil Indians Defend Sale of Gold, Trees. *Dallas Morning News*, November.

Errington, Shelly. 1993. Progressivist Stories and the Pre-Columbian Past. In *Collecting the Pre-Columbian Past: A Symposium at Dumbarton Oaks, 6th and 7th October 1990*, edited by Elizabeth Hill Boone. Washington: Dumbarton Oaks Research Library and Collection.

Ezzati, Majid, and Daniel M. Kammen. 2001. Indoor Air Pollution From Biomass Combustion and Acute Respiratory Infections in Kenya: An Exposure-Response Study. *Lancet* 358(9282), 25 August.

Fan, Xuetong, and Donald W. Thayer. 2001. Quality of Irradiated Alfalfa Sprouts. *Journal of Food Protection* 64(10):1574–1578, October.

Fantel, Hans. 1989. Sound: Christmas in Vienna with the First Nine Beethovens. *New York Times*, December 24.

————. 1990. Sound: Old Records Live by the Numbers—A CD Series is Notable for the Method It Uses in the Transfer of 78-rpm Originals. *New York Times*, May 6.

FAO (Commission on Genetic Resources for Food and Agriculture in Rome). 2001. Agreement Reached On Protecting Plant Genetic Resources: International Undertaking on Plant Genetic Resources, 16 July. http://www.fao.org.

FAO/IAEA. 2001. Information Sheet on International FAO/IAEA Symposium on the Use of Mutated Genes in Crop Improvement and Functional Genomics, Food and Agriculture Organization of the United Nations (FAO) and the International Atomic Energy Agency (IAEA), Vienna, Austria, 3–7 June 2002. http://www.iaea.org/worldatom/Meetings/Planned/2002/infcn89.shtml.

Faulder, Dominic. 1997. In the Name of Money: The SLORC, the Thais and Two Multi-national Oil Giants Are Building a Gas Pipeline. The Karen Are in the Way—and That's Too Bad. *Asiaweek* 23(18):42–43, 46–47, 9 May.

Feest, Christian F., ed. 1987. *Indians and Europe: An Interdisciplinary Collection of Essays*. Aachen: Edition Herodot.

————. 1990. Europe's Indians. In *The Invented Indian: Cultural Fictions and Government Policies*, edited by James A. Clifton. New Brunswick: Transaction Publishers.

Feiring, Alice. 2001. For Better or Worse, Winemakers Go High Tech. *New York Times*, 26 August.

Ferguson, Kirsty. 1997. Steiner's Philosophy on Compost: The Plot Thickens. *The Independent* (London), 1 November.

Ferguson, R. Brian. 1995. A Reputation for War. *Natural History* 104(4):62–63, April.

Feynman, Richard Phillips. 1964. *Lectures on Physics: Exercises*. Reading, MA: Addison-Wesley Pub. Co.

Financial Times. 2001. Pressure on the Pressure Groups. *Financial Times* (Editorial Comment), 13 July.

Fisher, Ian. 1999. In Congo War's Wake, a Massacre of the Wildlife. *New York Times,* 28 July.

Flannery, Timothy F. 1999. Debating Extinction. *Science* 283(5399):182–183, 8 January.

Fleming, Fergus. 2000. *Killing Dragons: The Conquest of the Alps.* New York: Atlantic Monthly Press.

Fletcher, June. 1998. Gear for the Ecologically Correct: A Few Green Gadgets Bring Some Environmentalism to Conventional Houses. *Wall Street Journal*, May 22.

Fogel, Robert W. 2000a. *The Fourth Great Awakening & The Future of Egalitarianism.* Chicago: The University of Chicago Press.

––––––. 2000b. The Extension of Life in Developed Countries and Its Implications for Social Policy in Twenty-first Century. *Population And Development Review* (Supplement), 26:291–317.

Fogel, Robert W., and Dora L. Costa. 1997. A Theory of Technophysio Evolution, With Some Implications for Forecasting Population, Health Care Costs, and Pension Costs. *Demography* 34(1):49–66, February.

Fox, Maggie. 1999a. Humans Wiped Out Australian Animals, Study Finds. *Reuters News Service,* 7 January. (Reprinted as Researchers Suspect Early Settlers Wiped Out Australia's Big Animals. *Houston Chronicle* 11 January.)

––––––. 1999b. Early Humans Left Trees for Dinner, Study Shows. *Reuters News Service*, *Nando Media* online, 14 January.

Fox, Richard G. 1969. Professional Primitives: Hunter-gatherers of Nuclear South Asia. *Man In India* 49(2):139–160, April–June.

Fox, Stephen. 1981. *John Muir and His Legacy: The American Conservation Movement.* Boston: Little, Brown & Co.

Frieden, Bernard V. 1979. *The Environmental Protection Hustle.* Cambridge, MA: The M.I.T. Press.

Friedman, Jonathon. 1994. *Cultural Identity and Global Process.* London: Sage.

Fry, Greg. 1997. Framing the Islands: Knowledge and Power in Changing Australian Images of 'the South Pacific.' *The Contemporary Pacific: A Journal of Island Affairs* 9(2):305–344, Fall.

Fumento, Michael. 1999. Truth Disrupters. *Forbes Magazine* 161(25):146–149, 16 November.

Furedi, Frank. 1999. Consuming Democracy: Activism, Elitism and Political Apathy. The European Science and Environment Forum online, 15 November.

Gamini, Gabriella. 1997. Sacked Crusader for Tribes Wins Body Shop Case. *Times* (London), Saturday, 7 June.

Gardner, John W. 1961. *Excellence: Can We Be Equal and Excellent Too?* New York: Harper.

Garn, Stanley M. 1994. Uses of the Past, *American Journal of Human Biology* 6(1):89–96.

Garofalo, Reebee. 1992a. Understanding Mega-Events: If We are the World, then How Do We Change It? In *Rockin' the Boat: Mass Music and Mass Movements,* edited by Reebee Garofalo. Boston: South End Press.

———. 1992b. Popular Music and the Civil Rights Movement. In *Rockin' the Boat: Mass Music and Mass Movements,* edited by Reebee Garofalo. Boston: South End Press.

Georgescu-Roegen, Nicholas. 1971. *The Entropy Law and The Economic Process.* Cambridge, MA: Harvard University Press.

Ghimire, Krishna B. 1994. Parks and People: Livelihood Issues in National Parks Management in Thailand and Madagascar. *Development and Change,* Guest editor: Dharam Ghai, 25(1):195–229, January. (Special Issue: Development and Environment: Sustaining People and Nature, Guest editor: Dharam Ghai.)

Gifford, Eli. 1998. *The Many Speeches of Seathl: The Manipulation of the Record on Behalf of Religious, Political and Environmental Causes.* Master's Thesis, Sonoma State University, n.d.

Gifford, Eli, and Michael R. Cook, eds. 1992. *How Can One Sell the Air?: Chief Seattle's Vision.* Summertown, TN: Book Publishing Co.

Gill, Sam. 1993. The Truth About Black Elk: A Biography Explores the Christianity of the Sioux Holy Man, review of *Black Elk: Holy Man of the Oglala,* by Michael F. Skeltenkamp. *New York Times Book Review* 34–35, 31 October.

Gimbutas, Marija. 1974. *The Gods and Goddesses of Old Europe: 7000 to 3500 B.C.: Myths, Legends and Cult Images.* Berkeley: University of California Press.

———. 1982. *The Gods and Goddesses of Old Europe, 6500 to 3500 B.C.: Myths and Cult Images.* Updated and retitled version of 1974 edition. London: Thames and Hudson.

———. 1989. *The Language of the Goddess.* With a foreword by Joseph Campbell. San Francisco and New York: Harper and Row.

Gober, Patricia. 1993. Americans on the Move. *Population Bulletin* 48(3):1–40, November.

Goodall, Jane. 1986. *The Chimpanzees of Gombe: Patterns of Behavior.* Cambridge, MA: Belknap Press of Harvard University Press.

———. 1990. *Through a Window: My Thirty Years with the Chimpanzees of Gombe.* Boston: Houghton Mifflin.

Gordon, Robert J. 1984. The !Kung in the Kalahari Exchange: An Ethnohistorical Perspective. In *Past and Present in Hunter Gatherer Studies,* edited by Carmel Schrire. Orlando: Academic Press.

———. 1986. End Note: A Namibian Perspective on Lorna Marshall's Ethnography. In *The Past and Future of !Kung Ethnography: Critical*

Reflections and Symbolic Perspective, Essays in Honor of Lorna Marshall, edited by Megan Biesele. Hamburg: Helmut Buske Verlag.

———. 1988. Bushman's Choice: Militarization or Plastic Stone Age. *Toward Freedom: Report on Non-alignment and the Developing Countries* 17–18, April/May.

———. 1990. Kicking Up a Kalahari Storm, a review of *Land Filled With Flies, A Political Economy of the Kalahari,* by Edwin N. Wilmsen. *Southern African Review of Books* 3(3 and 4):18–19, February, May.

———. 1992. *The Bushman Myth: The Making of a Namibian Underclass.* Boulder, CO: Westview Press.

———. 1999. Conserving the Bushman to Extinction. Preliminary paper, cited in Protected for the People or Against the People? *National Parks and Game Reserves in the Transvaal and Natal,* by Jane Carruthers. Perth, Western Australia: African Studies Centre of Western Australia, African Seminar Program, 5 February.

Gordon, Robert J., and Stuart Sholto Douglas. 2000. *The Bushman Myth: The Making of a Namibian Underclass.* 2nd ed. Boulder, CO: Westview Press.

Gough, Michael, and Steven J. Milloy. 1999. Dioxin in Ben & Jerry's Ice Cream, *Junkscience* online, 8 November.

Gowdy, John M. 1994. *Coevolutionary Economics: The Economy, Society, and the Environment.* Boston: Kluwer Academic Publishers.

———, ed. 1998. *Limited Wants, Unlimited Means: A Reader on Hunter-Gatherer Economies and the Environment.* Washington: Island Press.

Gowlett, John A. 2001. Archaeology: Out in the Cold, *Nature,* 413(6851): 33–34, 6 September.

Gray, Channing. 1997. You Just Can't Get Too Angry in a Tepee: People Escape From Modern Stress in Traditional Indian Dwellings. *Houston Chronicle* reprinted from the *Providence.* R.I. *Journal-Bulletin,* 9 July.

Grayson, Donald K. 1977. Pleistocene Avifaunas and the Overkill Hypothesis. *Science* 195(4279):691–692, 18 February.

———. 2001. The Archaeological Record of Human Impacts on Animal Populations. *Journal of World Prehistory* 15(1). Grzimek, Bernhard, and Michael Grzimek. 1961. *Serengeti Shall Not Die.* New York: Dutton.

Guha, Ramachandra. 1989. Radical American Environmentalism and Wilderness Preservation: A Third World Critique. *Environmental Ethics* 11(1), Spring.

———. 1997. The Authoritarian Biologist and the Arrogance of Anti-humanism: Wildlife Conservation in the Third World. *Ecologist* 27(1):14–20.

———. 1998. The Past and the Future of Indian Environmentalism. *Indian Horizons. Indo-Asian Culture* 45(3–4):161.

Guha, Sumit. 1998. Lower Strata, Older Races, and Aboriginal Peoples: Racial Anthropology and Mythical History Past and Present. *Journal of Asian Studies* 57(2) May.

Hames, Raymond. 1987. Game Conservation or Efficient Hunting? In *The Question of the Commons: The Culture and Ecology of Communal Resources,* edited by Bonnie M. McCay and James M. Acheson. Tucson: University of Arizona Press.

Hamilton, David B. 2001. Comment Provoked by Mason's "Dusenberry Contribution to Consumer Theory." *Journal of Economic Issues* 35(3):745-747, September.

Hammond, Sharon. 1997. White Hunters Join Up With Villagers. *African Eye/Misanet, Electronic Mail & Guardian,* online, 13 August.

———. 1999a. Mozambique Resumes Elephant Hunting. *African Eye News Service, Electronic Mail & Guardian,* online, 1 July.

———. 1999b. Africa's First Cross-Border Game Park. *Electronic Mail & Guardian* online, April.

Han, Kuo-huange, and Lindy Li Mark. 1980. Evolution and Revolution in Chinese Music. In *Musics of Many Cultures: An Introduction,* edited by Elizabeth May. Berkeley and Los Angeles: University of California Press.

Hanna, Mike. 1997. If It Pays, It Stays. *CNN* broadcast, 9 June.

Haraway, Donna. 1992. The Promise of Monsters. In *Cultural Studies,* edited by Lawrence Grossberg, Cary Nelson, and Paula Triechler. New York and London: Routledge.

Harmon, Amy. 2001. Suddenly, "Idea Wars" Take On a New Global Urgency. *New York Times,* 11 November.

Harper, Janice. 2002. *Endangered Species: Health, Illness and Death Among Madagascar's People of the Forest.* Durham: Carolina Academic Press (in press).

Harris-Jones, Peter. 1993. Between Science and Shamanism: The Advocacy of Environmentalism in Toronto. In *Environmentalist: The View from Anthropology,* edited by Kay Milton. London: Routledge.

Harrison, Gordon. 1968. Ecology: The New Great Chain of Being. *Natural History* 77(10):8–16, 60–67, December.

Havel, Vaclav. 1993. Revolutionary. Translated by Paul Wilson. *New Yorker,* 69(43):116, 20 December.

Hawken, Paul, Amory Lovins, and L. Hunter Lovins. 1999. *Natural Capitalism: Creating the Next Industrial Revolution.* Boston: Little, Brown and Co.

Headland, Thomas N., ed. 1992a. *The Tasaday Controversy: Assessing the Evidence.* Washington: American Anthropological Association, Scholarly Series No. 28.

———. 1992b. The Tasaday: A Hoax or Not? In *The Tasaday Controversy: Assessing the Evidence,* edited by Thomas N. Headland. Washington: American Anthropological Association, Scholarly Series No. 28.

Heazle, Michael. 1997. Business Aside: Green Way to Go. *Far Eastern Economic Review* 160(41):84, 9 October.

Hellerer, Ulrike, and K.S. Jarayaman. 2000. Greens Persuade Europe to Revoke Patent on Neem Tree. *Nature* 405(6784):266–267, 18 May.

Hellmich, Richard L., Blair D. Siegfried, Mark K. Sears, Diane E. Stanley-Horn, and Michael J. Daniels. 2001. Monarch Larvae Sensitivity to *Bacillus Thuringiensis*-purified Proteins And Pollen, Proceedings of the National Academy of Sciences USA 98(20):, 9 October, online 14 September.

Henige, David P. 1990. Their Numbers Became Thick: Native American Historical Demography as Expiation. In *The Invented Indian: Cultural Fictions and Government Policies*, edited by James A. Clifton. New Brunswick: Transaction Publishers.

————. 1998. *Numbers From Nowhere: The American Indian Contact Population Debate*. Norman, OK: University of Oklahoma Press.

Heuveline, Patrick. 1999. The Global and Regional Impact of Mortality and Fertility Transitions, 1950–2000. *Population and Development Review* 25(4):681–702, December.

Hightower, Elizabeth. 2001. Thin Air and Thick Men: A Review of *Killing Dragons: The Conquest of the Alps* by Fergus Fleming. *New York Times Book Review,* 7 January.

Hill, Kim, and A. Magdalena Hurtado. 1989. Observations of Foragers in Paraguay Suggest There is No Single Pattern of Hunter-Gatherer Behavior. *American Scientist* 77(5):436–444, September–October.

Hippocrates. 1989. Vital Statistics. *Hippocrates* 3(4):12, July/August.

Hiscoe, Helen B. 1983. Does Being Natural Make It Good? *The New England Journal of Medicine* 308(24):1474, 16 June.

Hitchcock, Robert K. 1993a. Africa and Discovery: Human Rights, Environment, and Development. *American Indian Culture and Research Journal* 17(1):129–152. (Special Edition: *International Year Of Indigenous Peoples: Discovery And Human Rights.*)

————. 1993b. Toward Self-Sufficiency. *Cultural Survival Quarterly* 17(2):51–53, Summer.

————. 1995. Centralization, Resource Depletion, and Coercive Conservation Among the Tyua of the Northeastern Kalahari. *Human Ecology* 23(2):169–198, June.

————, and Rodney Brandenburgh. 1990. Tourism, Conservation and Culture in the Kalahari Desert. *Cultural Survival Quarterly* 14(2):20–24.

————, and John D. Holm. 1993 Bureaucratic Domination of Hunter-Gatherer Societies: A Study of the San in Botswana. *Development and Change* 24(2):305–334, April.

Hochswender, Woody. 1990. The Green Movement in the Fashion World. *New York Times*. 25 March.

Hoggan, Karen. 2000. Neem Tree Patent Revoked: The Neem Provides a Popular Traditional Tooth Cleaner. *BBC World Service,* online, 11 May.

Holden, Constance. 1999. Random Samples: Urban Decay in Old Mexico. *Science* 283(5398):31, January.

Holliday, Sarah L., Alan J. Scouten, and Larry R. Beuchat. 2001. Efficacy of Chemical Treatments in Eliminating Salmonella and *Escherichia coli* O157:H7 on Scarified and Polished Alfalfa Seeds. *Journal of Food Protection* 64(10):1489–1495, October.

Holloway, Marguerite. 1993. Sustaining the Amazon. *Scientific American* 269(1):90–99, July.

Holt, Nancy. 1998. How 'Green' is Your Household?: New Earth-Friendly Homes Are Right Next Door; Straw Bales and Used Tires. *Wall Street Journal,* 22 May.

Homewood, Brian. 1995. Brazilian Court Bans Indians From 'Mining' Mahogany. *New Scientist* 147:5, 22 July.

Houston Press. 1990. This Land is Not for Sale. *Houston Press* 2(2):61, 24 May.

Howard, Alan. 1999. Pacific-Based Virtual Communities: Rotuma on the World Wide Web. *The Contemporary Pacific: A Journal of Island Affairs* 11(1):160–175, Spring.

Howard, Jennifer. 1997. The Poetry and Romance of Luxury Travel, a review of *Grand Tours and Cook's Tours: A History of Leisure Travel, 1750 to 1915,* by Lynne Withey. *Civilization: The Magazine of the Library of Congress* 4(1):81, February/March.

Howell, Nancy. 1976. Towards a Uniformitarian Theory of Human Paleodemography. In *The Demographic Evolution of Human Populations,* edited by Richard H. Ward and Kenneth M. Weiss. New York: Academic Press.

———. 1988. Images of Tasaday and the !Kung: Reassessing Isolated Hunters-Gatherers. Paper Presented at the Annual Meeting of the Society for American Archaeology. Tuscon, Arizona, April, 1988.

Huber, Toni. 1997. Green Tibetans. In *Tibetan Culture in the Diaspora: Papers Presented at a Panel of the 7th Seminar of the International Association for Tibetan Studies,* Graz, Austria, edited by Frank J. Korom. Wien: Verlag der sterreichischen Akademie der Wissenschaften.

Huget, Jennifer. 2001. Hold the Sprouts. *Washington Post,* 4 September.

Hulse, Joseph. 1982. Food Science and Nutrition: The Gulf Between Rich and Poor. *Science* 217(4552), 18 June.

Hunter, Malcolm L., Robert K. Hitchcock, Robert K. and Barbara Wyckoff-Baird. 1990. Women and Wildlife in Southern Africa. *Conservation Biology: The Journal of the Society for Conservation Biology* 4(4):448, December.

IATP (Institute for Agriculture and Trade Policy). 1995. 100% Cotton Tampons Arriving Soon. *Organic Cotton Monitor,* online 1(7), 31 October.

———. 1997a. European Union Approves Controversial Patent Directive. *Intellectual Property & Biodiversity News,* online, 6(10), 18 July.

———. 1997b. India Bio-piracy Campaign Successfully Challenges U.S. Tumeric Patent. *Intellectual Property & Biodiversity News,* online, 6(13), 22 September.

———. 1997c. News Briefs. *Labels: Linking Consumers and Producers,* online 1(6), 8 December.

ICL. International Constitutional Law Research Team. 1979. *A Report on Tribal Minorities in Mindanao.* Manila, RP: Regal Printing.

ICOMOS. International Council on Monuments and Sites. 1991. *Newsletter: International Committee on Cultural Tourism* 4(1).

Inambao, Chrispin. 1997. Impoverished San Trek to Angola. *The Namibian/Misa* on *Electronic Mail & Guardian,* online, 24 July.

Ingersoll, Bruce. 1989. Farming is Dangerous, but Fatalistic Farmers Oppose Safety Laws. *Wall Street Journal,* 29 July.

———. 2000b. *1999–2000 Annual Report of the International Rice Research Institute.* Los Banoe: International Rice Research Institute. 19 April.

Iovine, Julie V. 2001. Shades of Green: Muscle Houses Trying to Live Lean. *New York Times,* 30 August.

IRRI. 2000a. Rice Nations Urged to Prepare for PVP and IP. International Rice Research Institute. IRRI Press Release, 29 March.

Isenberg, Andrew C. 1993. Indians, Whites and the Buffalo: An Ecological History of the Great Plains, 1750–1900. Ann Arbor, MI: UMI Dissertation Services. PhD dissertation, Northwestern University.

———. 2000. The Destruction of the Buffalo: An Environmental History, 1720–1920. New York: Cambridge University Press.

Iwatani, Yukari. 1998. Battle Over Chinatown's Turtles, Frogs Reaches Boiling Point. Reuters News Service, *Nando Media,* online, 10 November.

Iyer, Pico. 2000. Room at the Top, review of *Life and Death on Mt. Everest: Sherpas and Himalayan Mountaineering,* by Sherry B. Ortner. *New York Review of Books* 46(16):22–24, 21 September.

Jackson, Jeremy B. C., Michael X. Kirby, Wolfgang H. Berger, Karen A. Bjorndal, Louis W. Botsford, Bruce J. Bourque, Roger H. Bradbury, Richard Cooke, Jon Erlandson, James A. Estes, Terence P. Hughes, Susan Kidwell, Carina B. Lange, Hunter S. Lenihan, John M. Pandolfi, Charles H. Peterson, Robert S. Steneck, Mia J. Tegner, and Robert R. Warner. 2001. Historical Overfishing and the Recent Collapse of Coastal Ecosystems. *Science* 293(5530):629–637, 27 July.

Jacoby, Karl. 2001. Crimes Against Nature: Squatters, Poachers, Thieves, and the Hidden History of American Conservation. Berkeley: University of California Press.

Jamieson, Neil L. 1985. Paradigms, Perceptions and Changing Reality. In *Cultural Values and Human Ecology in Southeast Asia*, edited by Karl L. Hutterer, A. Terry Rambo and George Lovelace. Ann Arbor: The University of Michigan, Center for South and Southeast Asian Studies, Michigan Papers on South and Southeast Asia.

———. 1993. *Understanding Vietnam*. Berkeley: University of California Press.

Jarayaman, K.S. 2000. As India Pushes Ahead with Plant Database. *Nature* 405(6784):267, 18 May.

Javna, John. 1990. 'Green' Message is Letter Well-sent. *Houston Chronicle*, 9 December.

Jenkins, Cathy. 1999. Gorilla Slaughter in Congo. *BBC World Service* online, 31 July.

———. 2000. Two-year Freeze on Ivory Trade: Kenya Wanted the Ivory Ban Restored. *BBC World Service* online, 17 April.

Jeursen, Belinda. 1996. Bushman Face Double Edge Sword, *Electronic Mail & Guardian* online, 5 September.

Johannes, R. E. 1978. Traditional Marine Conservation Methods in Oceania and Their Demise. *Annual Review of Ecology and Systematics* 9:349–364.

Johnson, George. 1995. Some Indians Buck a Stereotype. *New York Times* Section 4:6, Week in Review, 23 April.

Johnston, Barbara, and Ted Edwards. 1994. The Commodification of Mountaineering. *Annals of Tourism Research: A Social Science Journal* 21(3):459–478.

Johnston, David. 1993. Spiritual Seekers Borrow Indians' Ways. *New York Times* 147(49,558), 12 December.

Jones, Stephen, and Stephen Hallet. 1994. The Precious Traditions of Chinese Music. In *The Rough Guide to World Music* edited by Simon Broughton, Mark Ellingham, David Muddyman, and Richard Trillo, contributing editor, Kim Burton. London: Rough Guides Ltd.

Jordan, Charles, and Donald Snow. 1992. Diversification, Minorities, and the Mainstream Environmental Movement. In *Voices from the Environmental Movement: Perspectives for a New Era*, edited by Donald Snow. Washington, D.C.: Island Press for The Conservation Fund.

Jules-Rosette, Bennetta. 1984. *The Messages of Tourist Art: An African Semiotic System in Comparative Perspective*. New York: Plenum Press.

———. 1990. Simulations of Postmodernity: Images of Technology in African Tourist and Popular Art. *Society for Visual Anthropology Review* 6(1):29–37.

Juma, Calestous. 1989. *The Gene Hunters: Biotechnology and the Scramble for Seeds*. Princeton, NJ: Princeton University Press.

Kaiser, Rudolph. 1987. 'A Fifth Gospel, Almost:' Chief Seattle's Speech(es): American Origins and European Reception. In *Indians and Europe: An Interdisciplinary Collection of Essays,* edited by Christian F. Feest. Aachen, Germany: Edition Herodot.

KaKade, M.L., and I.E. Liener. 1973. The Increased Availability of Nutrients from Plant Foodstuffs Through Processing. In *Man, Food and Nutrition: Strategies and Technological Measures for Alleviating the World Food Problem,* edited by Miloslav Recheigl, Jr. Cleveland: CRC Press.

Kalland, Arne. 1994. Whose Whale is That? Diverting the Commodity Path. In *Elephants and Whales: Resources for Whom?*, edited by Milton M.R. Freeman and Urs P. Kreuter. Basel: Gordon and Breach Science Publishers.

Kamat, S.R. 1998. Air Pollution and Respiratory Problems. In *Respiratory Medicine in the Tropics*, edited by J.N. Pande. New Delhi: Oxford University Press.

Kamuaro, Ole. 1996. Ecotourism: Suicide or Development? *Voices from Africa,* Issue No. 6:59–65, August. (Special Issue on Sustainable Development.)

Kane, Joe. 1995. *Savages*. New York: Knopf; distributed by Random House.

Kasere, Stephen. 1996. Campfire: Zimbabwe's Tradition of Caring. *Voices from Africa,* Issue No. 6:33–39, August. (Special Issue on Sustainable Development.)

Keeley, Lawrence H. 1996. *War Before Civilization*. New York: Oxford University Press.

Keesing, Roger. 1990. Reply to Trask. *The Contemporary Pacific: A Journal of Island Affairs* 3(1):168–171, Spring.

———. 1993. Kastom Re-Examined. *Anthropological Forum* 6(4):587–596.

Kehoe, Alice. 1990. Primal Gaia: Primitivists and Plastic Medicine Men. In *The Invented Indian: Cultural Fictions and Government Policies,* edited by James A. Clifton. New Brunswick: Transaction Publishers.

Kennedy, Maev. 1999. Art News: Secrets of Museum Mummies. *Guardian* (London), 15 May.

Kenner, Hugh. 1987. *The Mechanical Muse*. New York: Oxford University Press.

Kerr, William A., Jill E. Hobbs, and Revadee Yampoin. 1999. Intellectual Property Protection, Biotechnology and Developing Countries: Will the Trips be Effective? *AgBioForum* 2(3 and 4):203–211.

Kirby, Alex. 2000. US Tree Patent Challenged: The Neem Provides a Popular Traditional Tooth Cleaner. *BBC World Service,* online, 5 May.

Kirch, Patrick V. 1979. Subsistence and Ecology. In *The Prehistory of Polynesia,* edited by Jesse D. Jennings. Cambridge, MA: Harvard University Press.

————. 1982. The Impact of the Prehistoric Polynesians on the Hawaiian Ecosystem. *Pacific Science* 36(1):1–14.

————. 1984. *The Evolution of the Polynesian Culture.* Cambridge: Cambridge University Press.

————. 1997a. Introduction: The Environmental History of Oceanic Islands. In *Historical Ecology in the Pacific Islands: Prehistoric Environmental and Landscape Change*, edited by Patrick V. Kirch and Terry L. Hunt. New Haven, CT: Yale University Press.

————. 1997b. Epilogue: Islands as Microcosms of Global Change? In *Historical Ecology in the Pacific Islands: Prehistoric Environmental and Landscape Change*, edited by Patrick V. Kirch and Terry L. Hunt. New Haven: Yale University Press.

Kirch, Patrick V., and Dana Lepofsky. 1993. Polynesian Irrigation: Archaeological and Linguistic Evidence for Origins and Development. *Asian Perspectives: The Journal of Archaeology for Asia and the Pacific* 32(2), Fall.

Kirch, Patrick V., and Terry L. Hunt, eds. 1997. *Historical Ecology in the Pacific Islands: Prehistoric Environmental and Landscape Change.* New Haven, CT: Yale University Press.

Kirshenblatt-Gimblett, Barbara. 1998. *Destination Culture: Tourism, Museums, and Heritage.* Berkeley: University of California Press.

Kjekshus, Helge. 1977. *Ecology Control and Economic Development in East African History: The Case of Tanganyika 1850–1950.* London: Heinemann.

Klein, Diane. 1993. When Crisis Hits, Cherished Possessions Come to Mind: Belongings Close to the Heart are Often Saved. *Houston Chronicle*, 30 October.

Klein, Richard. 1992. The Impact of Early People on the Environment: The Case of Large Mammal Extinctions. In *Human Impact on the Environment: Ancient Roots, Current Challenges*, edited by Judithe Jacobsen and John Firor. Boulder, CO: Westview Press.

Klieger, P. Christian. 1997. Shangri-La and Hyperreality: A Collision in Tibetan Refugee Expression. In *Tibetan Culture in the Diaspora: Papers Presented at a Panel of the 7th Seminar of the International Association for Tibetan Studies*, Graz, Austria, edited by Frank J. Korom. Wien: Verlag der sterreichischen Akademie der Wissenschaften.

Kloor, Keith. 2000. Restoration Ecology: Returning America's Forests to Their 'Natural' Roots. *Science* 287(5453):573–575, 28 January.

Knauf, Bruce M. 1987. Reconsidering Violence in Simple Human Societies: Homicide Among the Gebusi of New Guinea. *Current Anthropology* 28(4):457–500, August–October.

Knight, Danielle. 2000. Indian, Thai Farmers Fight US 'Biopiracy.' *Inter Press Service, Asia Times,* online, 2 May.

Knight, John. 1999. Monkeys on the Move: The Natural Symbolism of People-Macaque Conflict in Japan. *Journal of Asian Studies* 58(3):622–647, August.

Knox, Margaret. 1990. Africa Daze, Montana Knights: Wilderness As Armed Fortress. *Buzzworm: The Environment Journal* 2(4):46–51, July/August.

Koch, Eddie. 1995. Whose Land is This? *Electronic Mail & Guardian,* online, 23 February.

———. 1996. The Texan Who Plans a Dream Park Just Here. *Electronic Mail & Guardian,* online, January.

———. 1997. Playground Along the Beach of Poverty. *Electronic Mail & Guardian,* online, 6 May.

Koch, Eddie, David Cooper, and Henk Coetzee. 1990. *Water, Waste, and Wildlife: The Politics of Ecology in South Africa.* London & Johannesburg, South Africa: Penguin Books; distributed by Thorold's Africana Books.

Konner, Melvin. 1987. On Human Nature: False Idylls. *Sciences* 27(5):8–10, September/October.

Konner, Melvin, and Marjorie Shostak. 1986. Ethnographic Romanticism and the Idea of Human Nature: Parallels Between Samoa and !Kung San. In *The Past and Future of !Kung Ethnography: Critical Reflections and Symbolic Perspective, Essays in Honor of Lorna Marshall,* edited by Megan Biesele. Hamburg: Helmut Buske Verlag.

Koro, Emanuel, Juan Ovejero, and Julian Sturgeon. 1999. Hunters—'The Ultimate Ecotourists?' *The ACP Courier: Africa-Pacific-European Union,* No. 175:53–54, May–June.

Kozinn, Allan. 1999. Music Goes Down, Around and Back in Time. *New York Times,* 15 January.

Kraut, Alan M. 1994. *Silent Travelers: Germs, Genes and the 'Immigrant Menace'.* New York: Basic Books.

Krech, Shepard, III., ed. 1981. *Indians, Animals and the Fur Trade: A Critique of Keepers of the Game.* Athens, GA: The University of Georgia Press.

———. 1999a. *The Ecological Indian: Myth and History.* New York: W.W. Norton.

———. 1999b. Playing with Fire. *New Scientist* 164(2209):56–59, 23 October.

Kreuter, Urs P., and Randy T. Simmons. 1994. Economics, Politics and Controversy over African Elephant Conservation. In *Elephants and Whales: Resources for Whom?,* edited by Milton M.R. Freeman and Urs P. Kreuter. Basel, Switzerland: Gordon and Breach Science Publishers.

Lacey, Mark. 2001. Traditional Spirits Block a $500 Million Dam Plan in Uganda. *New York Times,* 13 September.

Laden, Francine, Gwen Collman, Kumiko Iwamoto, Anthony J. Alberg, Gertrud S. Berkowitz, Jo L. Freudenheim, Susan E. Hankinson, Kathy J.

Helzlsouer, Theodore R. Holford, Han-Yao Huang, Kirsten B. Moysich, John D. Tessari, Mary S. Wolff, Tongzhang Zheng, and David J. Hunter. 2001. Lysophosphatidic Acid Induction of Vascular Endothelial Growth Factor Expression in Human Ovarian Cancer Cells. *Journal of the National Cancer Institute* 93(10):768–775, 16 May.

LaFranchi, Howard. 1997. Amazon Indians ask 'Biopirates' to Pay Damages. *Christian Science Monitor, Nando.net,* online, 20 November.

Lamprey, Hugh. 1992. Challenges Facing Protected Area Management in Sub-Saharan Africa. In *Managing Protected Areas in Africa,* edited by Walter J. Lusigi. Paris: UNESCO-World Heritage Fund.

Landsburg, Steven E. 2001. The Imperialism of Compassion. *Wall Street Journal,* 23 July.

Leaver, Christopher J., and Anthony Trewavas. 2001. Comment on (Greenpeace's) Stefan Flothmann and Jan van Aken's Article "Of Maize and Men," *EMBO Reports* 2(9):744–745, September.

Lee, Richard B. 1969. Eating Christmas in the Kalahari. *Natural History* 78(10):14, 16, 18, 21–22, 60–63, December.

———. 1972. Population Growth and the Beginnings of Sedentary Life Among the !Kung Bushman. In *Population Growth: Anthropological Implications,* edited by Brian Spooner. Cambridge, MA: MIT Press.

———. 1979. *The !Kung San: Men, Women and Work in a Foraging Society.* Cambridge: Cambridge University Press.

———. 1992. Making Sense of the Tasaday: Three Discourses. In *The Tasaday Controversy: Assessing the Evidence,* edited by Thomas N. Headland. Washington: American Anthropological Association, Scholarly Series No. 28.

Lee, Richard B., and Irwen DeVore, eds. 1968. *Man the Hunter.* Chicago: Aldine.

———, eds. 1976. *Kalahari Hunter-Gatherers: Studies of the !Kung San and Their Neighbors.* Cambridge, MA: Harvard University Press.

Lencek, Lena, and Gideon Bosker. 1998. *The Beach: The History of Paradise on Earth.* New York: Viking.

Leonard, Neil. 1985. The Reactions to Ragtime. In *Ragtime: Its History, Composers and Music,* edited by John Edward Hasse. New York: McMillan Books.

Leopold, Aldo. 1933. *Game Management.* New York: C. Scribner's Sons.

———. 1966. *A Sand County Almanac: With Essays on Conservation from Round River.* New York: Oxford University Press, 1949 New York: Sierra Club/Ballantine.

Levy, Stuart B. 2001. Antibacterial Household Products: Cause for Concern. *Emerging Infectious Diseases Journal.* Paper presented at the International Conference on Emerging Infectious Diseases 2000 in Atlanta, Georgia 7(3) Supplement, June.

Line, Les. 1998. Birds in the Bush and the Database: The Gadgets Needed to Spot Birds Become Even More Sophisticated. *New York Times*, 25 June.

————. 1999. Indiana Jones Meets His Match in Burma Rabinowitz. *New York Times*, 3 August.

Lobe, Jim. 1999. Media Colonizes Image of Africa. *MISAnet/Inter Press Service, Woza,* online, 11 March.

Loesser, Arthur. 1954. *Men, Women and Pianos: A Social History.* New York: Simon and Schuster.

Lopez, Donald S. 1994. New Age Orientalism *Tricycle: The Buddhist Review* 36–43, Spring.

————. 1998. *Prisoners of Shangri-La: Tibetan Buddhism and the West.* Chicago: University of Chicago Press.

Lowenthal, David. 1985. The Past is a Foreign Country. Cambridge: Cambridge University Press.

Lutts, Ralph H. 1990. *The Nature Fakers: Wildlife, Science and Sentiment.* Golden, CO: Fulcrum Publishing.

Lyman, Francesca. 2000. Your Environment: Nature's Bedroom: Sweet Dreams for a Healthier You. *MSNBC,* online, 19 January.

Mabey, Richard. 1988. Flying into History, review of *Biographies for Birdwatchers: The Lives of Those Commemorated in Western Paleartic Bird Names,* by Barbara Mearns and Richard Mearns. *Nature* 336(6196):273, 17 November.

MacKenzie, John M. 1987. Chivalry, Social Darwinism and Ritualized Killing: The Hunting Ethos in Central Africa up to 1914. In *Conservation in Africa: People, Policies and Practice*, edited by David Anderson and Richard Gove. Cambridge: Cambridge University Press.

————. 1988. *The Empire of Nature: Hunting, Conservation and British Imperialism.* Manchester, England: Manchester University Press.

Macleod, Fiona. 1997a. Africa Must Pay for Its Wildlife: Legendary Kenyan Conservationist Richard Leakey Speaks in South Africa Against Private Game Reserves Run by Caucasians, and the Argument that Wildlife Needs to Pay Its Own Way. *Electronic Mail & Guardian,* online, 1 October.

————. 1997b. Seeing Red Among the Greens. *Electronic Mail & Guardian,* online, 14 November.

————. 1998. Bushmen Get a Stake in Game Park. *Electronic Mail & Guardian,* online, 23 December.

Maddocks, Melvin. 1989. What is This Thing Called Jazz? *World Monitor* 2(1):14–15, April.

Madeley, John. 2001. US Rice Group Wins Basmati Patents. *Financial Times*, 24 August.

Madison, Cathy. 1997. Green Death. *Utne Reader: The Best of The Alternative Media*, 84–85 September–October.

Makova, Patrice. 1997a. Zimbabwe Steps Up Fight for Ivory Trade. *PANA* (Pan-African News Agency), online, 3 March.

―――. 1997b. Too Many Jumbos Wreck the Parks. *PANA* (PanAfrican News Agency), online, 16 April.

Malan, Rian. 2000. In the Jungle. *Rolling Stone* No. 841:54–66, 84–85, 25 May.

Mann, Charles C. 2000a. Earthmovers of the Amazon. *Science* 287(5454):786–789, 4 February.

―――. 2000b. The Good Earth: Did People Improve the Amazon Basin? *Science* 287(5454):788, 4 February.

Manuel, Peter. 1988. *Popular Musics of the Non-Western World: An Introductory Statement.* New York: Oxford University Press.

―――. 1993. *Cassette Culture: Popular Music and Technology in North India.* Chicago: The University of Chicago Press.

Mapininga, Marvellous. 1997. Africa in Bid to Win Elephant Downlisting. *PANA* (PanAfrican News Agency), online, 8 June.

Marlar, Richard A., Banks L. Leonard, Brian R. Billman, Patricia M. Lambert, and Jennifer E. Marlar. 2000. Biochemical Evidence of Cannibalism at a Prehistoric Puebloan Site in Southwestern Colorado. *Nature* 407(6800):74–78, 7 September.

Marshall, Lorna. 1960. !Kung Bushman Bands. *Africa* 30(4): 325–355, October.

Martin, Calvin. 1979. *Keepers of the Game: Indian-Animal Relationship and the Fur Trade.* Berkeley: University of California Press.

―――. 1981. The American Indian as Miscast Ecologist. In *Ecological Consciousness: Essays from the Earthday X Colloquium,* edited by Robert C. Schultz and J. Donald Hughes. Washington: University Press of America. Paper presented at Earthday Colloquium, University of Denver, April 21–24, 1980.

Martin, Paul S. 1973. The Discovery of America: The First Americans May Have Swept the Western Hemisphere and Decimated Fauna Within 1000 Years. *Science* 179(4077):969–974, 9 March.

Martin, Paul S., and Richard G. Klein. eds. 1984. *Quaternary Extinctions: A Prehistoric Revolution.* Tucson: University of Arizona Press.

Martin, Paul S., and Christine R. Szuter. 1999. War Zones and Game Sinks in Lewis and Clark's West. *Conservation Biology: The Journal of the Society for Conservation Biology* 13(1):36–45, February.

Mashelkar, R. A. 2001. Our Patent Ignorance. *Times of India,* 24 August.

Matossian, Mary Kilbourne. 1989. *Poisons of the Past: Molds, Epidemics and History.* New Haven, CT: Yale University Press.

Matte, Thomas D., Michaeline Bresnahan, Melissa D. Begg, and Ezra Susser. 2001. Influence of Variation in Birth Weight Within Normal Range and Within Sibships on IQ at Age 7 Years: Cohort Study. *British Medical Journal* 323(7308):310–314, 11 August.

Matthiessen, Peter. 1995. Annals of Conservation: Survival of the Hunter. *New Yorker* 71(9):67–77, 24 April.

May, Sir Robert. 2001. Risk and Uncertainty: At the Frontiers of Science, We Don't Always Know What May Happen. *Nature* 411(6840):891, 21 June.

McCay, Bonne M., and James M. Acheson. eds. 1987. *The Question of the Commons: The Culture and Ecology of Communal Resources*. Tucson: University of Arizona Press.

McElroy, Ann, and Patricia K. Townsend. 1989. *Medical Anthropology in Ecological Perspective*. 2nd edition. Boulder, CO: Westview Press.

McKibben, Bill. 1989. *The End of Nature*. New York: Random House.

Meadows, Donella H., Dennis L. Meadows, Jorgen Randers, and William W. Behrens III. 1972. *The Limits to Growth: A Report for the Club of Rome's Project on the Predicament of Mankind*. New York: Universe Books.

Meadows, Donella H., Dennis L. Meadows, and Jorgen Randers. 1992. *Beyond the Limits: Confronting Global Collapse, Envisioning a Sustainable Future*. Mills, VT: Chelsea Green Pub.

Mearns, Barbara, and Richard Mearns. 1988. *Biographies for Birdwatchers: The Lives of Those Commemorated in Western Paleartic Bird Names*. New York: Academic Press.

Medawar, Jean, and David Pyke. 2000. *Hitler's Gift: Scientists Who Fled Nazi Germany*. London: Richard Cohen Books in association with the European Jewish Publication Society and Metro Publishing.

Medawar, Peter Brian. 1979. *Advice to a Young Scientist*. New York: Harper & Row.

———. 1984. *The Limits of Science*. New York: Harper & Row.

Mehra, Chander. 1999. Convention Allows Namibia, Zimbabwe to Sell Ivory. *Nation*. Nairobi, *Africa News,* online, 25 February.

Meister, C. 1976. Demographic Consequences of Euro-American Contact on Selected American Indian Populations to the Demographic Transition. *Ethnohistory* 23(2):161–172.

Meyer, Kathleen. 1989. *How to Shit in the Woods: An Environmentally Sound Approach to a Lost Art*. Berkeley: Ten Speed Press.

Michel, Karen Lincoln. 1995. Native Americans Upset at New Age Religious Interlopers: Some Say Sanctity of Rites Being Diminished. *Houston Chronicle*, 5 August.

Milbank, Dana. 1991. Being a Consumer Isn't Easy If You Boycott Everything. *Wall Street Journal* 137(80), 24 April.

Miller, Gifford H., John W. Magee, Beverly J. Johnson, Marilyn L. Fogel, Nigel A. Spooner, Malcolm T. McCulloch, and Linda K. Ayliffe. 1999. Pleistocene Extinction of *Genyornis Newtoni*: Human Impact on Australian Megafauna. *Science*, 283(5399):205–208, 8 January.

Mithen, Steven J. 1996. *The Prehistory of the Mind: The Cognitive Origins of Art, Religion and Science*. London: Thames & Hudson.

Moeller, Susan D. 1999. *Compassion Fatigue: How the Media Sell Disease, Famine, War, and Death*. New York: Routledge.

Moffett, Matt. 1994. Kayapo Indians Lose Their Green Image: Former Heroes of Amazon Succumb to Lure of Profit. *Wall Street Journal*, 224(126), 29 December.

Mohle-Boetani, J.C., J.A. Farrar, S.B. Werner, D. Minassian, R. Bryant, S. Abbott, L. Slutsker, and D.J. Vugia. 2001. *Escherichia coli* O157 and *Salmonella* Infections Associated with Sprouts in California, 1996–1998. *Annals of Internal Medicine* 135(4):239–247, 21 August.

Monbiot, George. 1994. *No Man's Land: An Investigative Journey Through Kenya and Tanzania*. London: Macmillan.

———. 1999. Whose Wildlife Is It Anyway, *Electronic Mail & Guardian*, online, 9 June.

Morgan, Stephen L. 2000. Richer and Taller: Stature and Living Standards in China, 1979–1995. *China Journal* (44):1–39, July.

Morris, Ramona, and Desmond Morris. 1966. *Men and Apes*. London: Hutchinson.

Mowat, Farley. 1987. *Women in the Mist: The Story of Dian Fossey and the Mountain Gorillas*. London: Macdonald & Co. Pub.

Mowforth, Martin, and Ian Munt. 1998. *Tourism and Sustainability: New Tourism in the Third World*. London: Routledge.

Mulenga, Mildred. 1997. African States Oppose Resumption of Ivory Trade, *PANA* (Panafrican News Agency), online, 20 May.

Murray, Gilbert. [1925] 1951. *Five Stages of Greek Religion*. Garden City, NY: Doubleday and Company, Inc.

Mutwira, Robin. 1989. Southern Rhodesian Wildlife Policy. 1890–1953: A Question of Condoning Game Slaughter? *Journal of Southern African Studies* 15(2):250–262 January. (Special Issue: *The Politics of Conservation in Southern Africa*).

Mwangi, George. 2000. Six Nations Reach Agreement on Elephants at Endangered Species Summit. *Associated Press*, *Nando Media*, online, 17 April.

Nagengast, Carole, and Terence Turner. 1997. Introduction: Universal Human Rights Versus Cultural Relativity. *Journal of Anthropological Research* 53(4):269–272.

Nanda, Meera. 1997. History Is What Hurts: A Materialist Feminist Perspective on the Green Revolution and Its Ecofeminist Critics. In *Materialist Feminism: A Reader in Class, Difference, and Women's Lives,* edited by Rosemary Hennessy and Chrys Ingraham. New York: Routledge.

———. 1998. The Episteme Charity of the Social Constructivist Critics of Science and Why the Third World Should Refuse the Offer. In *A House Built on Sand: Exposing Postmodernist Myths About Science,* edited by Noretta Koertge. New York: Oxford University Press.

Nandy, Ashis. 1988. The Human Factor. *Illustrated Weekly of India*, 17 January.

NAS. 1973. *Toxicants Occurring Naturally in Foods*. Washington: National Academy of Sciences, Committee on Food Protection, Food and Nutrition Board, National Research Council.

Nash, Jeffery E., and Anne Sutherland. 1991. The Moral Elevation of Animals: The Case of 'Gorillas in the Mist.' *International Journal of Politics, Culture and Society* 15(1):111–126.

Nash, Roderick Frazier. 1989. *The Rights of Nature*. Madison: The University of Wisconsin Press.

Nature. 2001. Italian Politicians Defend Pasta Source. *Nature* 411(6835):231, 17 May.

Navarro-Gonzalez, Rafael, Christopher P. McKay, and Delphine Nna Mvondo. 2001. A Possible Nitrogen Crisis for Archaean Life Due to Reduced Nitrogen Fixation By Lightning. *Nature* 412: 6842(61–64), 5 July.

Neale, Walter. 1973. Primitive Affluence. *Science* 179(4071):372–373, 26 January.

Nettl, Bruno. 1992. Heartland Excursions: Exercises in Musical Ethnography. *The World of Music: Journal of the International Institute for Traditional Music* 34(1):8–34.

Neumann, Roderick P. 1998. *Imposing Wilderness: Struggles Over Livelihood and Nature Preservation in Africa*. Berkeley: University of California Press.

New Scientist. 1995. Friendly Wool *New Scientist* 145(1970):13, 25 March.

New York Times. 1990. Eating to Save the Rain Forest. *New York Times*, 8 August.

New Yorker. 1990a. Goings on About Town: Night Life. *New Yorker*, 8–10, 9 July.

———. 1990b. Life in California. *New Yorker* 37, 9 July.

———. 1991a. Notes and Comment: The Talk of the Town. *New Yorker*, 29–30, 20 May.

———. 1991b. Note. *New Yorker* 116, 23 September.

Niewoehner, Wesley A. 2001. Behavioral Inferences From the Skhul/Qafzeh Early Modern Human Hand Remains. *Proceedings of the National Academy of Sciences of the United States of America* 98(6):2979–2984, 13 March.

Nichols, Rodney. 1997. Working Hypothesis: What If? *Sciences* 37(6):6, November/December.

Nicoll, Ruaridh. 1997. Would-be King Squeezes the Bushman. *Electronic Mail & Guardian*, online, 13 June.

NIEHS. 2001. DDT, PCBs Not Linked to Higher Rates of Breast Cancer, an Analysis of Five Northeast Studies Concludes, Washington, D.C.: National Institute of Environmental Health Sciences Press Release, NIEHS PR #01–13, 15 May.

North, Richard D. 1995a. End of the Green Crusade. *New Scientist* 145(1967):38–41, 4 March.

———. 1995b. *Life on a Modern Planet: A Manifesto for Progress*. New York: St. Martin's Press.

NRC. 1996. Committee on Comparative Toxicity of Naturally Occurring Carcinogens, Board on Environmental Studies and Toxicology, and the Commission on Life Sciences. National Research Council. *Carcinogens and Anticarcinogens in the Human Diet: A Comparison of Naturally Occurring and Synthetic Substances*. Washington, D.C.: National Academy Press.

Nyoni, Nadaba. 1997. Peasants Gain from Elephant Culling. *PANA* (Panafrican News Agency), online, 16 May.

Oates, Joyce Carol. 1973. New Heaven and Earth. *Arts in Society*, special issue, *The Humanist Alternative* 10 (1):36–43, Spring–Summer. First published in *Saturday Review/The Arts* 55(45):51–54, 4 November 1972.

Oberhauser, Karen S., Michelle D. Prysby, Heather R. Mattila, Diane E. Stanley-Horn, Mark K. Sears, Galen Dively, Eric Olson, John M. Pleasants, Wai-Ki F. Lam, and Richard L. Hellmich. 2001. Temporal And Spatial Overlap Between Monarch Larvae And Corn Pollen, Proceedings of the National Academy of Sciences, USA 98(20), 9 October. Online, 14 September.

Odhiambo, Nicodemus. 1999. Tanzania's Insensitive Tourism Abuses Ethnic Groups. *PANA* (Panafrican News Agency), 28 September.

Ogden, Michael. 1999. Island on the Internet. *The Contemporary Pacific: A Journal of Island Affairs* 11(2):451–462, Fall.

O'Loughlin, Ed. 1997. Bushmen Take a Feud to 20th-century Court. *Christian Science Monitor,* online, 22 November.

O'Neill, Molly. 1990. What to Put in the Pot: Cooks Face Challenge Over Animal Rights. *New York Times,* 8 August.

Ortega, Bob. 1995. The Bubbling Stream on Your Nature CD Might be a Toilet: Recording Artists Run the Gamut From Authentic to Ersatz; Are Surf Sounds All Alike? *Wall Street Journal* 96(12):1, A6, 19 July.

Ortner, Sherry B. 1999. *Life and Death on Mt. Everest: Sherpas and Himalayan Mountaineering*. Princeton, NJ: Princeton University Press.

Orwell, George. 1947. Lear, Tolstoy and the Fool. *Primitive* 8:2–17.

Ottaway, Marina. 2001. Reluctant Missionaries. *Foreign Policy*, July/August.

Palevitz, Barry A. 2001. When Science Gets in the Way of Pet Agendas. *Scientist*, 15(15):43, 23 July.

PANA: Panafrican News Agency. 1997. Zimbabweans Unhappy with Endangered Species Convention. *PANA*: Panafrican News Agency, online, 21 April.

Paredes, Oona Thommes. 1997. *People of the Hinterlands: Higaunon Life in Northern Mindanao, Philippines.* Master's Thesis, Arizona State University, n.d.

Pareles, Jon. 1989a. A Life of Giving Voice to Those Rarely Heard. *New York Times,* 7 March.

———. 1989b. In Music, 'Eurocentrism' Sparks a Vivacious Debate. *International Herald Tribune,* 28 April.

Parkes, Christopher. 2000. Letter from Los Angeles: No Alternative to the New Age Sales Pitch. *Financial Times,* London, 3 May.

Patterson, Orlando. 1994a. *Global Culture and the American Cosmos.* New York: Andy Warhol Foundation for the Visual Arts, Paper Series on the Arts, Culture, and Society, 2.

———. 1994b. Ecumenical America: Global Culture and the American Cosmos. *World Policy Journal* 11(3): Summer.

Pavlov, Pavel, John Inge Svendsen, and Svein Indrelids. 2001. Human Presence in the European Arctic Nearly 40,000 Years Ago. *Nature* 413(6851):64–67, 6 September.

Payne, Ronald. 1997. *Plus Ca Change*: Animal Rights and Anger in Ancient Culture. *European* 1–7 May.

Pearce, Fred. 1990a. The Green Missionaries of Africa. *New Scientist* 126(1713):64–65, 21 April.

———. 1990b. Bolivian Indians March to Save Their Homeland. *New Scientist* 127(1371):17, 25 August.

———. 1990c. *The Green Warriors.* London: Bodley Head.

———. 1997. Greenhouse Wars. *New Scientist* 155(2091):38–43, 19 July.

Peel, Quentin. 2001. NGOs Find Success Brings Problems. *Financial Times,* 12 July.

Perlez, Jane. 1991. Ramomafana Journal: Whose Forest Is It, the Peasants or the Lemurs? *New York Times,* 7 September.

Petean, Saulo. 1996. Broken Promises. *Brazzil,* 16–18, December.

Peterson, Dale, and Jane Goodall. 1993. *Visions of Caliban: On Chimpanzees and People.* Boston: Houghton Mifflin.

Peterson, John H. 1994. Sustainable Wildlife Use for Community Development in Zimbabwe. In *Elephants and Whales: Resources for Whom?,* edited by Milton M.R. Freeman and Urs P. Kreuter. Basel, Switzerland: Gordon and Breach Science Publishers.

Pleasants, John M., Richard L. Hellmich, Galen P. Dively, Mark K. Sears, Diane E. Stanley-Horn, Heather R. Mattila, John E. Foster, Thomas L. Clark, and Gretchen D. Jones. 2001. Corn Pollen Deposition on Milkweeds in And Near Cornfields, Proceedings of the National Academy of Sciences, USA 98(14), 9 October. Online, 14 September.

Pleydell-Bouverie, Jasper. 1994. Cotton without Chemicals. *New Scientist* 143(1944):25–29, 24 September.

Poirier, Marc R. 1996. Environmental Justice and the Beach Access Movement of the 1970s in Connecticut and New Jersey: Stories of Property and Civil Rights. *Connecticut Law Review* 28(719).

Poirine, Bernard. 1998. Should We Hate MIRAB? *The Contemporary Pacific: A Journal of Island Affairs* 10(1):65–105, Spring.

Pois, Robert A. 1986. *National Socialism and the Religion of Nature*. New York: St. Martin's Press.

Pollack, Andrew. 2001a. Data on Genetically Modified Corn Reports Say Threat to Monarch Butterflies Is "Negligible." *New York Times*, 8 September.

————. 2001b. New Research Fuels Debate Over Genetic Food Altering. *New York Times*, 9 September.

Pollitt, Katha. 1999. Father Knows Best. *Foreign Affairs* 78(1):122–125, January/February.

Poore, Patricia. 1989. Review of *How to Shit in the Woods: An Environmentally Sound Approach to a Lost Art,* by Kathleen Meyer. *Garbage* 1(1):58, September/October.

Porter, James. 1991. Muddying the Crystal Spring: From Idealism and Realism to Marxism in the Study of English and American Folk Song. In *Comparative Musicology and Anthropology of Music: Essays on the History of Ethnomusicology*, edited by Bruno Nettl and Philip V. Bohlman. Chicago: The University of Chicago Press.

Posey, Darrell. 1990. Intellectual Property Rights and Just Compensation for Indigenous Knowledge. *Anthropology Today* 6(4), August.

Postman, Lore. 1997. Baker's Goods Go to the Dogs: Canine Cafe Offers Cakes, Pizzas and Bottled Water. Charlotte Observer News Service, *Houston Chronicle,* 1 October.

Postrel, Virginia I. 2000. How Not to Treat Elephants Like Fish. *New York Times*, 18 May.

Postrel, Virginia. 2001. Criminalizing Science: Responses to a Left-right Alliance to Outlaw "Therapeutic Cloning" and Stigmatize Genetic Research. *Reason,* online, 21 October.

Powell, Mary Lucas. 1992. Health and Disease in the Late Prehistoric Southeast. In *Disease and Demography in the Americas*, edited by John W. Verano and Douglas H. Ubelaker. Washington: Smithsonian Institution Press.

Price, Sally. 1989. *Primitive Art in Civilized Places*. Chicago: The University of Chicago Press.

Proctor, Robert. 1995. *Cancer Wars: How Politics Shapes What We Know and Don't Know About Cancer*. New York: Basic Books.

Pye-Smith, Charlie. 1999. Truth Games: Is Our Desire to Save the World's Large Mammals Forcing Even the Most Dedicated Pressure Groups to Distort Data? *New Scientist* 161(2168):16–17, 9 January.

————. 2001. The Chainsaw's Last Stand, *Financial Times* (London), 11/12 August.

Pyne, Stephen J. 1982. *Fire in America: A Cultural History of Wildlife and Rural Fire*. Princeton, NJ: Princeton University Press.

————. 1991a. *Burning Bush: A Fire History of Australia*. New York: Henry Holt & Co., Inc.

————. 1991b. Fire Down Under: How the First Australians Put a Continent to the Torch. *Sciences* 31(2):39–45, March/April.

————. 1997. *Vestal Fire: An Environmental History, Told Through Fire, of Europe and Europe's Encounter with the World*. Seattle: University of Washington Press.

Rabinowitz, Alan. 1986. *Jaguar: Struggle and Triumph in the Jungles of Belize*. New York: Arbor House.

Radford, Tim. 1997. The Grinch that Stole Christmas Dinner. *Guardian*, (London) & Scripps Howard, *Nando.net*, online, 24 December.

Ramaphosa, Cyril. 1999. African Voices: Media for a New Age. Speech by Cyril Ramaphosa on the occasion of the CNN African Journalist of the Year Awards 1999, Johannesburg, South Africa, 18 March.

Rambo, A. Terry. 1985. *Primitive Polluters: Semang Impact on the Malaysian Tropical Rainforest Ecosystem*. Museum of Anthropology, Anthropological Paper No. 76. Ann Arbor: University of Michigan.

Rambo, A. Terry, Kathleen Gilloghy, and Karl. L. Hunter, eds. 1988. *Ethnic Diversity and the Control of Natural Resources in Southeast Asia*. Ann Arbor: University of Michigan, Center for South and Southeast Asian Studies, Michigan Papers in South and Southeast Asia, No. 32.

Ranger, Terence. 1989. Whose Heritage? The Case of Matobo National Park. *Journal of Southern African Studies* 15(2):217–249, January. (Special Issue: *The Politics of Conservation in Southern Africa*).

Ratliff, Ben. 2001. Jazz in the Catbird Seat: It Wasn't Always So. *New York Times*, 6 January.

Ravenhill, Philip L. 1995. Review of *African Art in Transit*, by Christopher B. Steiner. *African Arts* 28(2):16–17, Spring.

Reber, Pat. 1997. American Businessman Developing Huge Game Park in Mozambique. *Associated Press*, *Nando.net*, online, 9 December.

Recer, Paul. 2001. Damage to Earth Started in Ancient Times, Study Finds. *Associated Press*, *Nando Times*, online, 26 July.

Reeves, Tracey. 1997b. Tribe Opposes the Nomination of Clinton's Top Choice to Direct Indian Affairs. *Knight-Rider Tribune News*, *Houston Chronicle*, 8 November.

Reif, Wanda. 2001. Tragic Evolution of America's Indians: Native Land: Photographs from the Robert G Lewis Collection. *Lancet* 358(9279), 4 August.

Rembert, Tracey C. 1998. Natural Critter Care: Rethinking Food, Fun and Fleas. *E* IX(3):50–53, May–June.

Rensberger, Boyce. 1977. *The Cult of the Wild*. Garden City, NY: Anchor Press/Doubleday.

Reuters News. 2001. Turn Buried Bodies Into Organic Soil. *Reuters News Service,* online, 1 June.

Reyman, Theodore A., Hendrik Nielson, Ingolf Thuesen, Derek N. H. Notman, Karl J. Reinhard, Edmund Tapp, and Tony Waldron. 1998. New Investigative Techniques. In *Mummies, Disease, and Ancient Cultures*, edited by Aidan Cockburn, Eve Cockburn, and Theodore Allen Reyman. New York: Cambridge University Press.

Richer, David L. 2000. Intellectual Property Protection: Who Needs It? In *Agricultural Biotechnology and the Poor*, edited by Gabrielle J. Persley and M.M. Lantin. Washington: Consultative Group on International Agricultural Research and the US National Academy of Sciences.

RICPQL. 1996. *Caring for the Future: Making the Next Decades Provide a Life Worth Living: Report of the Independent Commission on Population and the Quality of Life*. New York: Oxford University Press.

Rifkin, Jeremy. 2001. This Is The Age Of Biology: Left And Right Are Finding Common Ground In Opposition To A Utilitarian View Of Life. *Guardian* (London), 28 July.

Rifkin, Jeremy with Ted Howard, Afterword by Nicholas Georgescu-Roegen. 1980. *Entropy: A New World View*. New York: Viking Press.

Ripin, Edwin M. 1988. History of the Piano. In *The New Grove Dictionary of Musical Instruments: Piano*, edited by Stanley Sadie. New York: W.W. Norton and Company.

Robbins, Jim. 1999. Historians Revisit Slaughter on the Plains. *New York Times*, 16 November.

Robinson, Geoffrey. 1995. *The Dark Side of Paradise: Political Violence in Bali*. Ithaca, NY: Cornell University Press.

Rockefeller, Lawrence. 1976. The Case for the Simple Life-Style. *Reader's Digest* III(2):61–65, February.

Rockwell, John. 1988. Leftists Causes? Rock Seconds These Emotions. *New Times*, 11 December.

———. 1994. The New Colossus: American Culture as Power Export. *New York Times*, Section 2, Arts and Leisure, 30 January.

Root, Deborah. 1996. *Cannibal Culture: Art, Appropriation, & Commodification of Difference*. Boulder: Westview Press.

Ross, Andrew. 1988. *No Respect: Intellectuals and Popular Culture*. New York: Routledge.

Ross, Doran H. 1992. Epilogue: The Future of Elephants, Real and Imagined. *Elephant: The Animal and Its Ivory in African Culture*, edited by Doran H. Ross. Los Angeles: Fowler Museum of Cultural History, University of California at Los Angeles.

Rostlund, E. 1957. The Myth of a Natural Prairie Belt in Alabama: An Historical Interpretation. *Annals of the Association of American Geographers*, 392–411.

Rothman, Stanley, and S. Robert Lichter. 1996. Is Environmental Cancer a Political Disease? In *The Flight From Science and Reason*, edited by Paul R. Gross, Norman Levitt, and Martin W. Lewis. New York: The New York Academy of Sciences.

Runnels, Curtis N. 1995. Environmental Degradation in Ancient Greece. *Scientific American* 272(3):96–99, March.

Ryback, Timothy W. 1990. *Rock Around the Bloc: A History of Rock Music in Eastern Europe and the Soviet Union*. New York: Oxford University Press.

Sahlins, Marshall D. 1972. *Stone Age Economics*. Chicago, IL: Aldine Atherton.

Sale, Kirkpatrick. 1990. *The Conquest of Paradise: Christopher Columbus and the Columbian Legacy*. New York: Knopf.

Sampat, Payal. 1998. Judgement Protects Indigenous Knowledge. *World Watch: Working for a Sustainable Future* 11(1):8, January/February.

Sanders, William T. 1992. Ecology and Cultural Syncretism in 16th-Century Mesoamerica. *Antiquity* 66(250):172–190, March.

Sanderson, A.T., and Edmund Tapp. 1998. Disease in Ancient Egypt. In *Mummies, Disease, and Ancient Cultures*, edited by Aidan Cockburn, Eve Cockburn, and Theodore Allen Reyman. New York: Cambridge University Press.

Saunders, Shelley R., Peter G. Ramsden, and D. Ann Herring. 1992. Transformation and Disease: Precontact Ontario Iroquoians. In *Disease and Demography in the Americas*, edited by John W. Verano and Douglas H. Ubelaker. Washington: Smithsonian Institution Press.

Sayagues, Mercedes. 1998. Mozambique's Not So Sugary Daddy. *Electronic Mail & Guardian,* online, 15 May.

———. 1999a. Death Doesn't Stop Coastal Development. *Electronic Mail & Guardian,* online, 18 May.

———. 1999b. Rupert May be Eyeing Mozambican Dream Park. *Electronic Mail & Guardian,* online, 8 June.

———. 1999c. Untouched Mozambique Coastal Region Under Threat. *Electronic Mail & Guardian,* online, 22 October.

———. 1999b. Maputo Reserve Hangs in the Balance. *Electronic Mail & Guardian,* online, 19 November.

Scarlett, Lynn. 1991. Make Your Environment Dirtier—Recycle. *Wall Street Journal*, 14 January.

———. 2000. "Doing More with Less:—Unsung Environmental Triumph?" In *Earth Report 2000: Revisiting the True State of the Planet*, edited by Ronald Bailey. New York: McGraw-Hill.

Scherreik, Susan. 1998. Investing on Faith. *Money Magazine,* online, 22 October.

Scheuplein, Robert J. 2000. Pesticides and Infant Risk: Is There a Need for an Additional Safety Margin? *Regulatory Toxicology and Pharmacology* 31(3):267–279, 1 June.

Schmetzer, Uli. 1999. Elephant's Plight Pits U.S. Elite Against Indians Hungry for Land. *Chicago Tribune,* online, 19 March.

Schmitt, Peter. 1969. *Back to Nature: The Arcadian Myth in Urban America.* New York: Oxford University Press. Reprinted Baltimore, MD: Johns Hopkins University Press.

Schoenbaum, David. 1998. When Violin Meets Violinist: The Player Calls the Tune, review of *The Violin Explained: Components, Mechanism and Sound* by Sir James Beament. *Financial Times,* London, 10–11 January.

Schrire, Carmel. 1980. An Inquiry Into the Evolutionary Status and Apparent Identity of the San Hunter-Gatherers. *Human Ecology* 8(1):9–32, March.

———, ed. 1984a. *Past and Present in Hunter-Gatherer Studies.* Orlando: Academic Press.

———. 1984b. Wild Surmises on Savage Thoughts. In *Past and Present in Hunter-Gatherer Studies,* edited by Carmel Schrire. Orlando: Academic Press.

———. 1995. *Digging Through Darkness: Chronicles of an Archaeologist.* Charlottesville: University Press of Virginia.

Schuller, Gunther. 1985. Rags, the Classics and Jazz. In *Ragtime: Its History, Composers and Music,* edited by John Edward Hasse. New York: McMillan Books.

———. 1986. *Musings: The Musical Worlds of Gunther Schuller.* New York: Oxford University Press.

———. 1989. *The Swing Era: The Development of Jazz, 1930–1945.* New York: Oxford University Press.

Schultz, Theodore W. 1965. *Transforming Traditional Agriculture.* New Haven, CT: Yale University Press.

Schuettler, Darren. 1998. Gold Barons Under Scrutiny for 'Apartheid Sins.' *Reuters News Service, Woza,* online, 11 November.

Schwadel, Francine. 1989. They Laughed at Brigham Young When He Chose Salt Lake City. *Wall Street Journal,* 25 September.

Sears, Mark K., Richard L. Hellmich, Diane E. Stanley-Horn, Karen S. Oberhauser, John M. Pleasants, Heather R. Mattila, Blair D. Siegfried, and Galen P. Dively. 2001. Impact of BT Corn Pollen on Monarch Butterfly Populations: A Risk Assessment, Proceedings of the National Academy of Sciences, USA 98(20), 9 October. Online, 14 September.

Sharp, John, and Stewart Douglas. 1996. Prisoners of Their Own Reputations? The Veterans of the 'Bushman' Battalions in South Africa. In *Miscast: Negotiating the Presence of the Bushman,* edited by Pippa Skotnes. Cape Town, South Africa: University of Cape Town Press for South African National Gallery.

Shelton, Anthony M., and Mark K. Sears. 2001. The Monarch Butterfly Controversy: Scientific Interpretations of a Phenomenon. *Plant Journal* 27(6):483-488, October.

Shepherd, Steven L. 2000. The Mysterious Technology of the Violin. *American Heritage of Invention & Technology* 15(4):26–31, 34–37, Spring.

Shi, David E. 1985. *The Simple Life: Plain Living and High Thinking in American Culture*. New York: Oxford University Press.

Shostak, Marjorie. 1981. *Nisa: The Life and Words of a !Kung Woman*. Cambridge, MA.: Harvard University Press.

Showers, Kate B. 1994. The Ivory Story, Africans and Africanists. *Issue: A Journal of Opinion* 23:41–46, Winter/Spring.

Sithole, Emelia. 1999. San Get New Lease for Survival. *Reuters, Woza,* online, 23 March.

Skeltenkamp, Michael F. 1993. *Black Elk: Holy Man of the Oglala*. Norman: University of Oklahoma Press.

Skotnes, Pippa, ed. 1996. *Miscast: Negotiating the Presence of the Bushman*. Cape Town, South Africa: University of Cape Town Press for South African National Gallery.

Slesin, Susan. 1989. Back to Nature ... Well Not Really. *New York Times*, 14 September.

Smil, Vaclav. 2000. Rocky Mountain Visions: A Review Essay of Hawken, Lovins and Lovins. *Population and Development Review* 26(1):163–176, March.

———. 2001. *Enriching the Earth: Fritz Haber, Carl Bosch, and the Transformation of World Food*. Cambridge, MA: MIT Press.

Smith, Kirk R. 1983. Biomass Fuels, Air Pollution and Health. *Ambio: A Journal of the Human Environment* XIV(4–5):26.

Sokal, Alan. 1996a. Transgressing the Boundaries: Towards Transformative Hermeneutics of Quantum Gravity. *Social Text* (46/47):217–252, Spring/Summer.

———. 1996b. A Physicist Experiments with Cultural Studies. *Lingua Franca* 62–64, May/June.

———. 1998. What the Social Text Affair Does and Does Not Prove. In *A House Built on Sand: Exposing Postmodernist Myths About Science*, edited by Noretta Koertge. New York: Oxford University Press.

———, and Jean Bricmont. 1998. *Fashionable Nonsense: Postmodern Intellectuals' Abuse of Science*. New York: Picador.

Solman, Paul. 1999. World Bank Doubts Prices Recovery. *Financial Times*, London, 3 February.

Solomon, Enver. 1998. Green Activists Target Thai-Burma Pipeline. *BBC World Service,* online, 9 January.

Spence, Mark. 1996. Dispossessing the Wilderness: Yosemite Indians and the National Park Ideal, 1846 to 1926. *Pacific Historical Review* 115(1):27–59.

Spoonheimer Matt, and Julia A. Lee-Thorp. 1999. Isotopic Evidence for the Diet of an Early Hominid, *Australopithecus africanus. Science* 283(5400):368–379, 15 January.

Spriggs, Mathew. 1997. Landscape Catastrophe and Landscape Enhancement: Are Either or Both True in the Pacific? In *Historical Ecology in the Pacific Islands: Prehistoric Environmental and Landscape Change*, edited by Patrick V. Kirch and Terry L. Hunt. New Haven, CT: Yale University Press.

Stackhouse, John. 1996. Village Flirtation with Business Falls Flat: Grief in Ghana. *Toronto Globe and Mail*, 3 January.

Stanley-Horn, Diane E., Galen P. Dively, Richard L. Hellmich, Heather R. Mattila, Mark K. Sears, Robyn Rose, Laura C. H. Jesse, John E. Losey, John J. Obrycki, and Les Lewis. 2001. Assessing the Impact of Cry1AB-expressing Corn Pollen on Monarch Butterfly Larvae in Field Studies. 2001. Proceedings of the National Academy of Sciences USA 98(20), 9 October. Online, 14 September.

Stansfield, Sally, and Donald S. Shepard. 1993. Acute Respiratory Infection. In *Disease Control Priorities in Developing Countries*, edited by Dean T. Jamison, W. Henry Mosley, Anthony R. Meacham, and Jose Luis Babadilla. New York: Oxford University Press for the World Bank.

Steadman, David W. 1995. Prehistoric Extinctions of Pacific Island Birds: Biodiversity Meets Zooarchaeology. *Science* 267(5201):1123–1131, 24 February.

———. 1997. Extinctions of Polynesian Birds: Reciprocal Impacts of Birds and People. In *Historical Ecology in the Pacific Islands: Prehistoric Environmental and Landscape Change*, edited by Patrick V. Kirch and Terry L. Hunt. New Haven, CT: Yale University Press.

Steiner, Christopher B. 1994. *African Art in Transit*. Cambridge: Cambridge University Press.

Steiner, Rudolf. 1958. *Agriculture: A Course of Eight Lectures*. Translated by George Adams. London: Biodynamics Association.

Steinfels, Peter. 1990. Idyllic Theory of Goddess Creates Storm: Was a Peaceful Matriarchal World Shattered by Patriarchal Invaders? *New York Times*, 19 February.

Steinhart, Edward. 1994. National Parks and Anti-Poaching in Kenya, 1947–1957. *International Journal of African Historical Studies* 27(1):59–76.

Stern, Cassandra. 1998. Dispute Over Selling Live Animals as Food Erupts in Chinatown. *Washington Post,* news service, *Houston Chronicle*, 16 December.

Stevens, William K. 1993. An Eden in Ancient America?: Early Farming Left a Heavy Mark on the Landscape. *New York Times*, 30 March.

———. 1999. Unlikely Tool for Species Preservation: Warfare. *New York Times*, 30 March.

Stevenson, Robert L. 1994. *Global Communication in the Twenty-first Century*. New York: Longman.

Stevenson, Robert L., and Stille, Alexander. 2002. *The Future of the Past*. New York: Farrar, Straus & Giroux (in press).

Stock, Jon. 1999. 'Patent Buccaneers' Targeted. *South China Morning Post*, Internet Edition, 1 December.

Stocks, Anthony. 1987. Resource Management in an Amazon Varzea Lake Ecosystem: The Cocamilla Case. In *The Question of the Commons: The Culture and Ecology of Communal Resources*, edited by Bonnie M. McCay and James M. Acheson. Tucson, AZ: University of Arizona Press.

Storey, Rebecca. 1998. Mortality Through Time in a Poor Residence of the Pre-Industrial City of Teotihuacan. American Anthropological Association Meeting, Philadelphia, December.

Street, Brian. 1992. British Popular Anthropology: Exhibiting and Photographing the Other. In *Anthropology and Photography 1860–1920*, edited by Elizabeth Edwards. New Haven, CT: Yale University Press in association with The Royal Anthropological Institute.

Stuart, Anthony J. 1991. Mammalian Extinctions in the Late Pleistocene of Northern Eurasia and North America. *Biological Reviews of the Cambridge Philosophical Society* 66(4):453–563, November.

Sullivan, Andrew. 1999. Counter Culture: The Assault on Good News. *New York Times Magazine*, 7 November.

Survival International. 1991. Tourism: Special Issue, *Survival* 28.

———. 1994. Survival International's Contacts with the Body Shop. Paper Prepared at the Request of the Ethical Investment Research Services. EIRIS. London: Survival International, October.

———. 1995. Tourism and Tribal People. Background Sheet. London: Survival International.

Sutherland, Anne. 1996. Tourism and the Human Mosaic in Belize. *Urban Anthropology* 25(3):259–281, Fall.

———. 1998. *The Making of Belize: Globalization in the Margins*. Westport: Bergin & Garvey.

———, and Jeffery E. Nash. 1994. Animal Rights as a New Environmental Cosmology. *Qualitative Sociology* Special Issue: Animals in Social Relations 17(2):171–186, Summer.

Tata, Padma. 2001. Basmati Victory for India. *New Scientist* 171(2306):13, 1 September.

Tauxe, Robert V. 2001. Food Safety and Irradiation: Protecting the Public from Foodborne Infections. *Emerging Infectious Diseases Journal* (International Conference on Emerging Infectious Diseases 2000: Presentation from the 2000 Emerging Infectious Diseases Conference in Atlanta, Georgia) 7(3) Supplement, June. (Presented at the

International Conference on Emerging Infectious Diseases 2000 in Atlanta, Georgia).

Taylor, Dorcetta. 1992. Can the Environmental Movement Attract and Maintain the Support of Minorities? In *Race and the Incidence of Environmental Hazards: A Time for Discourse*, edited by Bunyan Bryant and Paul Mohai, pp. 28–54. Boulder, CO: Westview Press.

Taylor, Russell D. 1994. Elephant Management in the Nyaminyami District, Zimbabwe. In *Elephants and Whales: Resources for Whom?*, edited by Milton M.R. Freeman and Urs P. Kreuter, pp. 113–127. Basel, Switzerland: Gordon and Breach Science Publishers.

Teaford, Mark F., and Peter S. Ungar. 2000. Diet and the Evolution of the Earliest Human Ancestors. Proceedings of the National Academy of Sciences of the United States of America 97(25):13506–13511, 5 December.

Teleki, Geza. 1973. *The Predatory Behavior of Wild Chimpanzees.* Lewisburg, PA: Bucknell University Press.

———. 1981. The Omnivorous Diet and Eclectic Feeding Habits of Chimpanzees in Gombe National Park, Tanzania. In *Omnivorous Primates: Hunting and Gathering in Human Evolution,* edited by Robert S.O. Harding and Geza Teleki. New York: Columbia University Press.

Thomas, Cynthia. 1993. Perfume Scent-strip Makers Sniff at Success. *Houston Chronicle*, 8 July.

Thomas, Elizabeth Marshall. 1959. *The Harmless People.* New York: Vintage Books.

Thomas, Keith. 1983. *Man and the Natural World: A History of Modern Sensibility.* New York: Pantheon.

Thomas, Nicholas. 1994. *Colonialism's Culture: Anthropology, Travel and Government.* Princeton, NJ: Princeton University Press.

Thomas, Stephen J. 1994. Seeking Equity in Common Property Wildlife in Zimbabwe. In *Elephants and Whales: Resources for Whom?*, edited by Milton M.R. Freeman and Urs P. Kreuter. Basel, Switzerland: Gordon and Breach Science Publishers.

Thurow, Roger. 1989. For the Bushman, It's Not the Gods That Must Be Crazy. *Wall Street Journal* 84(8), 13 July.

Tiger, Lionel. 1987. *The Manufacture of Evil: Ethics, Evolution and the Industrial System.* New York: Harper & Row.

Tobias, Michael. 1998. The New Population Bomb: An Interview With Michael Tobias. *Mother Earth News* 48–53, August/September.

Toby, Sidney. 2000. Acid Test Finally Wiped Out Vitalism, And Yet.... *Nature* 408(6814):767, 14 December.

Tomlinson, John. 1991. *Cultural Imperialism.* Baltimore: Johns Hopkins University Press.

Tompkins, Jane P. 1992. *West of Everything: The Inner Life of Westerns.* New York: Oxford University Press.

Torrence, Robin, ed. 1989. *Time, Energy and Stone Tools*. New York: Cambridge University Press.

Trask, Haunani-Kay. 1991. Natives and Anthropologist: The Colonial Struggle. *The Contemporary Pacific: A Journal of Island Affairs* 3(1):159–167, Spring.

Trigger, Bruce G. 1981. Ontario Native People and the Epidemic of 1634–1640. In *Indians, Animals and the Fur Trade: A Critique of Keepers of the Game*, edited by Shepard Krech III, pp. 39–58. Athens, GA: The University of Georgia Press.

Trimble, Joseph E. 1988. Stereotypical Images, American Indians, and Prejudice. In *Eliminating Racism: Profiles in Controversy*, edited by Phyllis A. Katz and Dalmas A. Taylor. New York: Plenum Press.

Turner, Frederick. 1977. *The North American Indian Reader*. New York: Penguin Books.

Turner, James West. 1997. Continuity and Constraint: Reconstructing the Concept of Tradition from a Pacific Perspective. *The Contemporary Pacific: A Journal of Island Affairs* 9(2):345–381, Fall.

Turner, Terence. 1991. Representing, Resisting, Rethinking: Historical Transformations of Kayapo Culture and Anthropological Consciousness. In *Colonial Situations: Essays on the Contextualization of Ethnographic Knowledge*, edited by George W. Stocking, Jr.. Madison, WI: University of Wisconsin Press.

———. 1993a. The Role of Indigenous Peoples in the Environmental Crisis: The Example of the Kayapo of the Brazilian Amazon. *Perspectives in Biology and Medicine* 36(3):526–545, Spring.

———. 1993. Anthropology and Muticulturalism: What is Anthropology That Multiculturalists Should Be Mindful of It? *Cultural Anthropology: Journal of the Society for Cultural Anthropology* 8(4):411–429, November.

———. 1997b. Human Rights, Human Difference: Anthropology's Contribution to an Emancipatory Cultural Politics. *Journal of Anthropological Research* 53(4):273–291.

Turner, Terence, and Davi Kopenawa Yanomami. 1991. I Fight because I am Alive. *Cultural Survival Quarterly* 15(3):59–64, Summer.

Twain, Mark. [1874] 1911. Enchantments and Enchanters. In *Life on the Mississippi,* by Mark Twain. New York: Harper and Brothers.

Tyler, Patrick E. 1995. For Rockers, China is a Hard Place: Rock-and-Roll's Dwindling Fans Lament the Lost Days of Rebellion. *New York Times*, 9 January.

Ubelaker, Douglas H., and John W. Verano. 1992a. Introduction. In *Disease and Demography in the Americas*, edited by John W. Verano and Douglas H. Ubelaker. Washington: Smithsonian Institution Press.

———. 1992b. Conclusion. In *Disease and Demography in the Americas*, edited by John W. Verano and Douglas H. Ubelaker. Washington: Smithsonian Institution Press.

Udall, Stewart. 1972. The Indians: The First Americans, First Ecologists. In *Look to the Mountain Top*, Robert L. Iacopi with Bernard L. Fontana and Charles Jones. San Jose: Gousha Publications.

UNDP (United Nations Development Programme). 2001. *Human Development Report 2001: Making New Technologies Work For Human Development*. New York: Oxford University Press, Published for the United Nations Development Programme (UNDP).

Uscher, Nancy. 1990. Peter Sculthorpe: Responding to Nature. *Strings* V(3):49–51, November/December.

Uys, Cheryl. 1998. TRC to Probe War Atrocities Against San. *Electronic Mail & Guardian,* online, 3 July.

Vanderwerth, W.C. 1971. *Indian Oratory: Famous Speeches by Noted Indian Chieftains*. Norman: University of Oklahoma Press.

Veblen, Thorstein. 1922. *The Instinct of Workmanship and the State of Industrial Arts*. New York: B. W. Huebsch.

Verhovek, Sam Howe. 2001. Radical Animal Rights Groups Step Up Protests. *New York Times*, 11 November.

Vogel, Gretchen. 1999. Did Early African Hominids Eat Meat? *Science* 283(5400):303, 15 January.

Vollers, Maryanne. 1987. The Rhino Wars: Zimbabwe is Shooting Poachers Who Menace the Rare Black Rhino. *Sports Illustrated* 66(9):61–72, 2 March.

Walker, Kaetheryn. 1998. *Homeopathic First Aid for Animals: Tales and Techniques for a Country Practitioner*. Rochester, VT: Healing Arts Press.

Ward, R. Gerald. 1993. South Pacific Island Futures: Paradise, Prosperity, or Pauperism? *The Contemporary Pacific: A Journal of Island Affairs* 5(1):1–21, Spring.

Warren, Louis S. 1997. *The Hunter's Game: Poachers and Conservationists in Twentieth-Century America*. New Haven, CT: Yale University Press.

Weatherford, Jack McIver. 1991. *Native Roots: How the Indians Nourished America*. New York: Crown Books.

Weber, Eugen. 1994. Naiads in Flannel: A Review of Alain Corbin, *The Lure of the Sea: The Discovery of Seaside in the Western World, 1750–1840. The Times Literary Supplement* 4759, 17 June.

Welker, Robert Henry. 1955. *Birds and Men: American Birds in Science, Art, Literature and Conservation, 1800–1900*. Cambridge, MA: Harvard University Press.

Wells, Ken. 1997. Animal Farm: African Game Ranchers Seek a New Way to Save Endangered Species. *Wall Street Journal,* 7 January.

West, Elliott. 1995. *The Way to the West: Essays on the Central Plains*. Albuquerque: University of New Mexico Press.

Whelan, Elizabeth M. 1993. *Toxic Terror: The Truth Behind the Cancer Scares*. New York: Prometheus Books.

White, Robert. 1990. *Tribal Assets: The Rebirth of Native America*. New York: Henry Holt.

Wilford, John Noble. 1999. Study of Prehumans' Teeth Suggests That They Dined on Meat: Hints of High-Protein Animal Food Even Before Stone Tools. *New York Times*, 15 January.

————. 2001. New Evidence of Early Humans Unearthed in Russia's North. *New York Times*, 6 September.

Wilkinson, Todd. 1999. Charges of Racial Insensitivity Beset Environmentalists. *Christian Science Monitor*, 23 November.

Wille, Chris. 1991. Peace is Hell. *Audubon: Magazine of the National Audubon Society* 93(1):62–70, January.

Williams, Raymond. 1973. *The Country and the City*. New York: Oxford University Press.

Willis, David. 2000. Sweat Sells: Ancient Ritual Draws Tourist Dollars, Activist Outrage. *Civilization: The Magazine of the Library of Congress* 7(2):23–24, April/May.

Wilmsen, Edwin N. 1983. The Ecology of Illusion: Anthropology and Foraging in the Kalahari. *Reviews in Anthropology* 10:9–20, Winter.

————. 1989. *Land Filled With Flies: A Political Economy of the Kalahari*. Chicago: The University of Chicago Press.

————. 1990. Comment on Jacqueline S. Solway and Richard B. Lee, Foragers, Genuine or Spurious? Situating the Kalahari San in History. *Current Anthropology* 31(2):136–137, April.

————. 1999. Personal Email Communication, July.

Wilmsen, Edwin N., and J. Denbow. 1991. Paradigmatic History of San-Speaking Peoples and Current Attempts at Revision. *Current Anthroplogy* 31(5):489–524, December.

Wilson, P. S. 1992. What Chief Seattle Said. *Environmental Law* 22(4):1,451–1,468.

Wilson, Richard. 1999. Ivory Ban Lifted: The Tusks Can Raise Money for Conservation. *BBC World Service,* online, 10 February.

Withey, Lynne. 1997. *Grand Tours and Cook's Tours: A History of Leisure Travel, 1750 to 1915*. New York: Morrow.

Wolcott, James. 1993. On Television: P.C. or Not P.C. *New Yorker* 69(29):124–126, 13 September.

Woodard, Colin. 1999. U.S. Patents on Living Things Irk Some. *Christian Science Monitor, Nando Media,* online, 1 August.

Worsdale, Andrew. 1996. Comic Fantasist Jamie Uys Bows Out. *Electronic Mail & Guardian,* online, 2 February.

Wrangham, Richard W., and Dale Peterson. 1996. *Demonic Males: Apes and the Origins of Human Violence*. Boston: Houghton Mifflin.

Yeld, John. 1999. Joy as Displaced Desert People Get Land Back. *The Cape Argus, Africa News,* online, 22 March.

Yengoyan, Aram. 1991. Shaping and Reshaping the Tasaday: A Question of Cultural Identity—a Review Article. *The Journal of Asian Studies* 50(3):565–573, August.

ZA. 1999. ZA * NOW NEWS: They Said It. *Electronic Mail & Guardian,* online, 5 July.

Zangerl, A. R., D. McKenna, C. L. Wraight, M. Carroll, P. Ficarello, R. Warner, and M. R. Berenbaum. 2001. Effects of Exposure to Event 176 *Bacillus Thuringiensis* Corn Pollen on Monarch And Black Swallowtail Caterpillars Under Field Conditions, Proceedings of the National Academy of Sciences, USA 98(20), 9 October. Online, 14 September.

Zerner, Charles. 1996. Telling Stories About Biological Diversity. In *Valuing Local Knowledge: Indigenous People and Intellectual Property Rights*, edited by Stephen B. Brush and Doreen Stabinsky. Washington: Island Press.

Zerzan, John. 1998. Future Primitive. In *Limited Wants, Unlimited Means: A Reader on Hunter-Gatherer Economics and the Environment*, edited by John Gowdy, foreword by Richard Lee. Washington: Island Press.

Zimmermann, Erich W. 1951. *World Resources and Industries: A Functional Appraisal of the Availability of Agricultural and Industrial Materials.* New York: Harper & Brothers.

Zubok, Vladislav. 1998. Cold War Postscript. *CNN*, 19 October.

Index